FIRST PRINCIPLES

FIRST PRINCIPLES

A Primer of Ideas for the College-Bound Student

by
HUGH HEWITT

REGNERY GATEWAY
Washington, D.C.

Library of Congress Cataloging-in-Publication Data
Hewitt, Hugh, 1956-
First principles.

1. College student orientation—United States. 2. Ideology. 3. Thought and thinking. I. Title
LB2343.32.H49 1987 378'.198 87-4559
ISBN 0-89526-793-4

Published in the United States by
Regnery Gateway
1130 17th Street, NW
Washington, D.C. 20036

Distributed to the trade by
Kampmann & Company, Inc.
9 E 40th Street
New York NY 10016

10 9 8 7 6 5 4 3 2 1

For
Betsy

Contents

A NOTE TO PARENTS

This is a book about ideas. If you are considering purchasing it, you have already cleared most of the obstacles on the road to your student's college education. The Stanley Kaplan courses and the SATs are behind you. More recently, you have been urging your kid to get on with the applications, to write the essays and send the papers away, certified mail, of course. Only April 15 and the acceptance/rejection results await, along with the bills.

But it is all ahead of him or her. And most of what lies ahead will be far, far beyond your control, as it probably should be.

There are a few more things you can do to prepare them. One is to buy this book and ask them to read it. It is designed as an intellectual shield from bad ideas and bad choices. It is written to be readable. It is relentlessly chatty and casual in tone because it has been my experience with high school students that they appreciate this approach.

The chapters are not cleverly titled. You do not have to

read the book to figure out what is going on inside. A glimpse at the Table of Contents spells it out. This is a modest attempt to reach a few students, no more than two or three in any given high school, who could become truly talented, creative thinkers.

It is not a self-help book. It is not full of psychobabble. There is more sex on The Cosby Show than there is within. It is different from 95% of the books they buy for themselves.

Those who are bound for college need a book about ideas, because college will bombard students with ideas from day one. One idea will be to join the Hare Krishna. Another will be to accept the Malthusian dilemma as an explanation for the woes of Central America. Another will be the argument that cocaine, as a non-addictive substance with a powerful kick, is the drug of choice. Still another will be that a particular course of study leading inexorably to a particular profession is the correct one. There are more ideas than record albums on most campuses.

Your sons and daughters are in no danger if they are the dense sort, content to plod along, grab decent grades, grab beers, and eventually grab a diploma. But if they appear at all interested in puzzling things out, be prepared for some major changes in their world view. Even if they are diplomatic with you, the serious student is going to wonder how you ever came to believe what you believe.

This development is a good thing, and eventually their experiments with ideas will be repaid in spades. You may still be, I think should be, concerned about what ideas, convictions, and principles they will pick over the next four years. If you are, you can still try to lay some foundation for that intellectual growth.

No great promises are made. This book sets out some of the big, recurring issues and describes some of the central disagreements that spark the controversies surrounding them. A few facts, some anecdotes, and a lot of what appears to me to be logic is mixed into the discussions. I hope to avoid the description "dull."

You may want to read this book first and make margin notes for your student along the lines: "This is a boldface lie," or, "Ask me about this when you get a chance." I have always enjoyed reading the annotated copies of other people's books. That way I get not only the author's views but another reader's reactions as well. I also believe that your student is far more likely to read something when you have left behind a set of comments. There is the danger that the student will hold you accountable for what you write and may even prove adept at refuting you. You might enjoy that though.

At bottom, this book is an attempt to encourage deliberate and critical thinking on some key ideas and to fill in some of the astonishing gaps in knowledge I have found common in students just finishing high school or just beginning college. Before long, these students will be weighed down by detail and specialty. This book is for the generalist before he or she becomes a specialist. It is breezy, relatively short, and moderately priced. Invitations to join the intellectual fray do not often come from parents. Surprise your child with your confidence in his or her ability to read, understand, and enjoy a book of this sort.

INTRODUCTION

This book is intended for a limited but recurring audience. It is a book the likes of which I would have found extremely useful about a decade ago. Around that time, October of 1973, I had finished college applications to Harvard, Princeton, Cornell, and Denison. The SATs were history. The first semester courses of my senior year barely held my attention and would not have done so at all but for the vague threat that the grades received therein might figure in the decisions of the various Admissions Committees. The book is also fairly short and, I hope, entertaining. It carries none of the burdens of Eliot's *Silas Marner*.

If you have gone so far as to pick up this book and read more than the dust jacket—which may or may not have something to do with the content of the book, depending upon the occult practices of my publisher—you are an ambitious American. Most Americans do not read non-fiction books on what appear at first glance to be rather dense and difficult subjects. Most Americans prefer to get their non-fiction information from sources such as newspapers and TV that make gentler demands on the attention span. You have taken an additional step. But you are used to additional steps. It takes additional steps to gain the admiration of your classmates and win through to the top post of the student government. It takes additional steps to win or even compete in the Westinghouse science competition, or edit the school underground paper, or engage in the fierce rat-a-tat-tat of fact spitting that passes for debate competition. Additional steps to learn and somewhat master an instrument so that admirers wander away from a concert remarking on your "gift" when that gift is in fact the hard-earned result of thousands of hours of patient practice. Additional steps to walk your church's corridors and teach the third-graders their elementary Bible lessons. Additional steps to take courses at the local junior college or, if you are a preppie, to seek out something more intellectually rigorous than your typical fare.

So you are an ambitious American. But young. Seventeen or eighteen is not an age of any great enlightenment. For twelve years you have been playing the chords of intellectual growth, learning the basic arrangements. The next four years, the college years, will refine these talents for many of you. Some of you will then seek additional education and others will go to work. Very, very few of you will even then have begun to assemble a set of first principles. Nevertheless, your mind is about to come under assault.

Between now and your graduation from college you will listen to hundreds of voices—the voices of students, professors, and academic advisors. Further input will come from books, periodicals, and news reports. All of these voices will share a common aim: To influence your way of thinking. If even one of them tells you this, I will be surprised. Professors teach for money, but of the many I have known, all have also taken up teaching in the hope of attracting adherents to their views. Fellow students have views as well. They too will want you to be convinced of the correctness of those views. Newspapers and newsmakers want your attention for a variety of reasons, but to attract and hold that attention they will have to get a grappling hook into your mind.

None of these people will "show you the ball." In Washington, the term "hidden agenda" is currently hot. The hidden agenda is the set of goals that any particular person carries around in his head. All of that person's public life is devoted to gaining these goals. All of his various activities can be fairly described as tactics to achieve his goals. To display the hidden agenda too early is to risk an unnecessary rejection of it, so it remains largely hidden.

Recall your current or most recent affair of the heart. What is the goal of that affair? Let me assume it is the noble goal of a mutually satisfying affection, call it reciprocated love. When you first fixed your desire on your opposite number, did you then and there walk over and announce: "I intend for you to fall in love with me, that is my goal!"? Of course not. Not even if you are the worst sort of incompetent at such things. Rather, you began to plot.

A lover's plot can be as complicated as the most elaborate maze. It will involve all sorts of tactical decisions: Where you sit in the cafeteria. When you arrive at the football game. Where you park in the school lot. What courses you take. When and if the first few tactics pay off and dating begins, new tactics evolve. And on it goes. As Frank Herbert repeats frequently in his magnificent novel of science fiction, *Dune:* "plans within plans within plans."

Hidden agendas are a part of everyday life. People want to sell you their own ideas. But the hard sell infrequently if ever succeeds. So roundabout the seller comes, feeding you this morsel and this bit and this piece. Eventually you will have swallowed the whole thing. You may not be able to digest it and you may toss it back. But it will have had the chance to take hold. Vendors of ideas need nothing more than that chance.

When I was a sophomore in college I had an afternoon section in a course devoted to comparative European politics. "Sections" are small classes of 10 -15 undergraduates usually led by a graduate student, and are combined with lecture hours given by a professor whose audience can reach into the hundreds. The lectures and the readings are analyzed and discussed in the sections. Often, the real nuts and bolts of learning are accomplished there, and the section leader typically grades the papers and exams.

This particular section was led by Joel, a rotund, red-faced, genuinely likeable grad student. The topic was the development of industrial Britain and the political change that accompanied that development. As Joel recounted his view of this period of history, a conclusion formed in my mind. He kept mentioning things like "the critical accumulation of capital." He frequently referred to the "labor element" and the slow maturation of this element's consciousness. He spent a lot of time on the Chartist movement. A few more buzz-words surfaced and suddenly I realized that Joel was a Marxist. I was genuinely surprised to meet a Marxist much less be taught by one, and I blurted out, "But that's Marxist!" Ten or so heads swiveled to better view the idiot in their midst while Joel just smiled and acknowledged that his view was the Marxist view and that he was, in fact, a Marxist.

Joel was an excellent section leader, well-prepared for class and usually accessible outside of it. He was not a Stalinist, but had spent a couple of summers in Cuba and had come away impressed with Castro's society. All of these terms like Marxist and Stalinist may not mean anything to you yet. No matter. What matters is that Joel had led the section for a few weeks and had never stated that he was a Marxist. But as any good Marxist will tell you, Marxism affects and informs every opinion of its adherents. Joel was putting a Marxist spin on his teaching without alerting us to that crucial fact. He had hidden the ball.

Marxist academics are not the only ones guilty of violating truth-in-labelling standards. Perhaps you have heard of Lyndon Larouche, a three-time candidate for President who has often been described as "far right" by the media. Larouche's supporters are found in the lobbies of nearly every major American airport, selling books and magazines and soliciting contributions. Because many Americans have become very suspicious of Larouche's intentions and would not give a nickle or a minute to one of his followers, these airport predators do not attract customers by yelling out "Support Larouche!" Instead, they call out such questions as "Do you favor a strong national defense?" and thereby reel-in the unsuspecting passerby.

Marxists, Larouchites, scientific advocates of natural selection theory--all are adherents to certain specific sets of ideas. So are resolute Christians, Moslems, and Jews. So are New Right Republicans and NOW feminists. So is Earl Weaver. So am I, and so you will become. The question is, will you know what you are when you get there? The second question is, will you have ever made a conscious choice along the way?

My friend Nancy is a senior at the University of Michigan. In the summer of 1985, she was an "intern" in the Washington office of United States Senator Carl Levin. Before her stint on the Hill, she was an intern in the editorial office of the national lobbying group, Common Cause.

Nancy is probably a lot like my typical reader, though a shade older. She is earnest, smart, and a talented writer on the U. of M. paper, *The Michigan Daily.* As a high school student, she helped her class accomplish the list of things most classes do: run a student government, put on an annual play, keep the tennis team competitive. She even got involved in a fairly nasty wrangle with the local school board over that body's sudden reintroduction of final exams for seniors. She acquitted herself well in all these things. You probably recognize something of yourself in this brief resume.

When she arrived at college she possessed a keen mind, and many attitudes. But attitudes—such as "It's important to keep up with the news"—are very different from ideas. Ideas are the first step on the road to first principles. Ideas that really persuade an individual harden into convictions. Convictions that mature into guideposts for personal behavior and belief rise to the level of first principles. Attitudes are way down below ideas.

So Nancy had many attitudes and some ideas. She may also have been in the process of developing some convictions. When she arrived in Washington, however, she proclaimed an intention to "keep an open mind." This announcement resulted from my hectoring her about the ideological predispositions of the senator and the lobby for whom she would soon be working. Common Cause especially alarmed me because it parades as a "public interest" group, a "citizens' lobby." These are verbal facades very useful to the whole business of hiding the ball. Common Cause routinely, indeed almost inevitably, takes positions that are also adopted by various members of the left-wing of the Democratic Party. Fine. The battle of ideas requires committed contestants from all sides.

But Common Cause does not give the following speech to incoming interns; at least nothing like this was heard by Nancy:

> Welcome to Common Cause. We are a lobbying group. We call ourselves a public interest group, but our mem-

> bership is really only a fraction of other lobbying groups popularly identified as right-wing, such as the National Rifle Association. Still, the description of public interest group helps us with fund-raising, so we hang on to it. We take positions on a host of issues. These positions coincide with those of the left-wing of the Democratic Party in most cases. We hope you leave here believing as we believe, and to help ensure that result, we won't be exposing you to any articulate spokesmen of positions commonly identified as conservative. Enjoy your stay.

Such a speech would be an up-front warning of the intellectual spin that interns would be receiving over the summer. Perhaps Common Cause expects its interns to know this, but I doubt it.

So Nancy arrived declaring "an open mind," but went to work in places sealed off from certain sets of ideas. If, day after day, you are fed only French cuisine, you will eventually learn to love French cuisine. You may not lose your taste for other types of cooking, but if a discussion turns to the subject of cooking and the relative merits of different styles of food preparation, you will talk about French cuisine because you know something about it. If you are like most people you will also come to defend this style of cooking as the best available because it is the style you are accustomed to and have learned to savor. Had you already been a gourmet of other cuisines before your steady diet of French cooking began, you might still argue for the superiority of French-style preparation, but your judgment would not only be more credible, it would also probably display more balance, temperance, and a willingness to acknowledge the best points of the competitors' claims.

When the summer of 1985 closed, Nancy finished a three month period marked by a steady absorption of ideas that are, broadly speaking, leftist. Her intellectual diet was uniform. She heard the same opinions on many occasions, and she may have come to repeat them herself. Had she arrived at the door of Common Cause with a firmly established set of first principles, she would no doubt still have found many of the positions of that special interest group to be appealing. But some would not have been palatable. Some would have had to be rejected as incompatible with her independently-arrived-at convictions.

Nancy has not, of course, been brainwashed. She is, rather, a victim of selective exposure. That selective exposure will mark her for some time, perhaps a lifetime. There is nothing wrong with this result. But there is something preferable to it: The conscious choice of personal convictions from among a true variety of possibilities.

This book is about choosing your own set of political and intellectual principles. It sets out the basic outlines of the big issues of the late Twentieth century. These are issues that fairly command the attention of every American who exercises the right to vote. They are not that difficult to understand, and intelligent choice among positions is actually pretty easy. Consciously or unconsciously, every American who votes does make a choice on most of these issues, although a couple are non-political. You who read this book are probably going to want to do a great deal more than simply vote. Most of you will want to argue about such things as nuclear freeze movements, unemployment levels, the need for wilderness areas, the existence of God, the ethics of abortion, and a thousand other things upon which you will be called to have an opinion. Some of you will go beyond discussion into political activity in support of various causes. A few may even seek elective or appointive office.

To involve yourself in such discussions without adequate preparation is to risk violating Swift's maxim that "It is better to remain silent and be thought a fool than to open one's mouth and remove all doubt." There is an alternative to remaining silent or exposing your own foolishness. The alternative is to be thoughtful and thus to sound thoughtful. This is a difficult goal to reach. Much more difficult than excelling in athletics or art. But it is attainable if you take it in steps. Once you have established in your own mind a few concrete convictions and a set of first principles, you will have laid the foundation for a thoughtful and perhaps even wise approach to political and moral problems. No one will be able to hide the ball from you for long.

MONEY

A subject both practical and theoretical is a good one on which to launch a series of discussions. Money inspires many emotions, particularly greed. As a college student it will be difficult for you to focus on much other than the pressing *need* for money that the great majority of you will feel at every turn. Try, however, to delay this concern and concentrate for a few pages on some general ideas about money.

Ideas about money abound: How to get it, how to use it, how to steal it without getting caught. There are also numerous technical questions, such as whether "the money supply" regulated by the Federal Reserve Board is on target. But for a general introduction to the subject, only two basic questions matter. First, is money a good thing? Second, if it is a good thing, what general conditions will lead to its increase?

As to the first question, please spare me the answer, "It depends." While it most certainly does "depend" on thousands of factors, it will depend on thousands of factors you will never build into your answer. You must instead pon-

der the question awhile and then decide whether, on the whole, money is a good thing or an evil thing.

There are plenty of straightforward answers to this question. Sophocles was one of the great playwrights of the ancient world. In one of his plays, *Antigone,* appears this conclusion:

> Money lays waste to cities; it sets men to roaming from home; it seduces and corrupts men and turns virtue to baseness; it teaches villainy and impiety.

A more traditionally American view is found in the mouth of Poor Richard, Benjamin Franklin's creation:

> 'Nothing but money is sweeter than honey."

Obviously, there are two sides here.

Consider this scenario: Ten dollars can purchase a sizeable amount of marijuana. Ten people purchasing ten bags each at this price over the course of a year will funnel $1,000 to the small- time seller who lives on the block. If that seller keeps half of the take, and along with nine other local merchants, sends the other half to the primary supplier in the city, that supplier will have $5,000 on hand. Ten town suppliers pay off the regional boss who pockets $25,000, and the ten regions in the one quadrant of the state produce for the quadrant chief a personal take of $125,000. The person running the state network picks up a half-million, and his boss is sure to make double that amount. The boss pays tribute to one of the dope overlords of Mexico who payrolls a small army of thugs. One of these thugs helps kidnap, torture, and murder an officer of the United States Drug Enforcement Agency in Mexico, as happened in late 1984. As that thug takes a cab home later in the day of the murder, he pulls out a ten and pays off the driver.

It is your ten dollar bill, spent a year ago. Money, even small amounts of money, can have enormous consequences. Is money a good thing?

You are watching the Live Aid Concert. Phil Collins is excellent. A number flashes on the screen. You dial it and pledge ten dollars. You copy down the address given over the phone and send off your money. Five million similar souls do the same thing, and, three months later, 30 trucks depart for remote areas of the Sudan, where the margin of life is narrow indeed. Your ten dollars has purchased several pounds of rice that feeds for a week a seven-year-old who weighs 20 pounds. Another person's ten dollars takes over after that. Money, even small amounts of money, can have enormous consequences.

The temptation grows to say "It depends." But these two examples show that the answer does not "depend," but is unambiguously "Yes." Consider that without the private donations sparked by Live Aid, many thousands of people would have died, but that the DEA agent did not have to die as an inevitable consequence of money's presence in the world. Money did not trigger the agent's execution. Had the people in original possession of the money chosen not to purchase drugs, there would have been no drug network, and no thugs. It was possible to have both a live agent and a live child. But if money had simply not existed, there would be no dead agent but most certainly a dead child.

Because money can do good things, it is a good thing. It can also be, but is not necessarily, an evil thing. Thus, a first principle becomes clear that is not so much a conviction as a fact: The world needs money in order to do good things. Money funds hospitals and parks, purchases off-road vehicles, provides the research that finds cures for disease, allows weekends at the shore and the acquisition of books. A relatively huge amount of it will buy you what passes for an education. Without money, a great many good things cannot be done. Because of money, nothing evil must necessarily—inevitably—happen. So question one has a definite answer: Money is a good thing.

Armed with this conclusion, a corollary also becomes clear. The world, as a whole, needs as much money as possible. Why? Because more and more money means more and more of the good things that make life happy: more rice for refugees, more sailboats, more ramps for those in wheelchairs, more summer homes on the Cape. But what conditions produce more money, by which I mean, economic growth? Economic growth depends on many factors, including the availability of natural resources and the advance of science and technology. Perhaps the most crucial factor is the "economic system" that different governments around the world adopt. There are three models.

The first model is one in which the government of a country owns the means of production—it owns the farms, factories, and stores. In the second model, the government leaves business pretty much alone. The first model is called socialism and is marked by "central planning"; the government decides what to make, where to build it, etc. The second is the "free market", and there is as little planning by government officials as possible. Individuals and individual corporations make economic decisions. The third model is the "regulated economy", where the government sets rules that govern economic activity, and individuals are free to make money any way they can so long as they do not violate these rules. America has gradually moved from the second, "free market" model to the third, "regulated" model. The first model has been tried in various places for various lengths of time. It has never come close to producing as much economic growth as either of the other two. Few people dispute this. Many people prefer socialism, but none can honestly say that it produces more economic growth than the other two models. That is, quite simply, a lie.

I do not intend to catalogue the economic failures of socialism. They are too many and too obvious to deserve review. But the choice between a regulated economy and a free market economy is a very difficult one, and there are hardly any advocates of a pure tilt one way or another. I would like to illustrate why the debate is so lively, and why sometimes more regulation—more rules on the ways you can make money—is preferred and why sometimes it is not.

Since the industrial revolution began there has been a huge demand for coal, which has led to a large expansion in mining activity. Coal mine owners must make a profit. If they do not, their mines shut down. Before the government began to pass laws telling coal mine owners that they had to operate their mines according to such-and-such rules, the coal mines made more money because their costs were lower. After the laws passed, mine owners had to do various things such as install ventilation equipment and safety devices, and they also had to clean up an area after they were finished strip-mining it. Since all these things cost money, mine owners faced a choice: accept less profit or raise the price of coal.

Without the ventilation equipment and the safety devices, many coal miners develop lung disease or suffer disabling accidents. Perhaps before it passed the new laws, the government reviewed the data, checked the bottom line, and decided that while the new laws would certainly require costly changes in coal mining, the changes in the long-run would save the lives and the money of mine workers whose health was imperiled.

Most likely, the government did not get too sophisticated in its economic analysis, but rather surveyed the high number of dead and disabled coal miners and decided that the level of suffering was too high. No doubt someone stood on the floor of Congress and said, "You cannot put a price on a human life."

Perhaps you cannot. But perhaps you can and should.

When I was in law school, I had a professor, Joseph Sax, who was extraordinarily bright. He is and has been for years on the leading edge of the environmentalist movement. Students generally thought him "a liberal," but he was an honest liberal, as was illustrated by his lecture on "the boy-in-the-well" dilemma.

Take the situation of the boy who has fallen into a well, his lecture began. The local authorities spare no expense to get the boy out of the well. They will spend thousands and thousands of dollars to save the boy.

But, Professor Sax pointed out, the same city officials will often refuse to spend money on obvious improvements in public safety. Perhaps they balk at buying a second emergency vehicle and staffing it with paramedics, even though that project would cost about the same as saving the boy in the well, and would probably save many, many more lives.

Sax's point was that the government responded without regard to cost to the drama of the boy in the well, while the less dramatic, but every bit as real need for a second ambulance was decided on the issue of cost alone, even though the objective of both situations was the same: saving life. Government is revealed as a less than completely rational decision-maker, which is not surprising because government is simply a collection of less than rational men and women. The boy-in-the-well dilemma is only a simple example of the "costs and benefits" analysis that affects most regulatory decisions. A more sophisticated example is found in the daily work of the Occupational Safety and Health Administration ("OSHA"). OSHA routinely regulates the air pollution levels companies can have inside their factory walls. To meet this "standard" the companies must often spend huge amounts of money. If OSHA determines that particular gas particles are so dangerous that none of them can be allowed in the factory air, the company must pay for the equipment to keep them out. Presumably, the number of workers who get sick because of the gas particles in the air will go down. Maybe. OSHA almost never knows for sure because with most of the gas particles the Agency studies, there are no definite answers on how dangerous the particles are in small amounts. As OSHA decides on how stringent a standard to set, the companies produce estimates of how much the different standards would cost. Lost long ago was the idea that a worker might be willing to accept the risks of more particles in the air if that risk increased the company's profit and translated into a boost in wages.

The rub is that most regulatory questions boil down to a comparison of how much money the regulation will cost and how much protection that regulation will produce. If the government demands too much protection in the workplace, for example, a factory might have to shut down. Workers will lose their jobs and the product—suppose it is fertilizer that Live Aid wants to purchase and ship to the Sudan—is not made anymore.

That is all pretty theoretical. Try a different, less exotic example.

Take McDonald's. Surely you or someone you know works at McDonald's. If they work around the French fry baskets, they are bound to get burned occasionally by splattering grease, but usually not seriously. Assume the risk is there, however, that someone could slip and put his hand into that mass of boiling oil, possibly losing the use of that hand forever.

What is the government to do? Should it order McDonald's to develop a device that keeps the grease covered almost all of the time? Then if a person slips, he or she will have his hand hit the cover rather than plunge into the grease.

But the cover costs a lot. To put it over all the fry pits and keep it there in working order will cost, say, $5,000 a year. Suppose a single McDonald's sells 50,000 bags of French fries a year. To pay for the new cover, the price of fries is increased a dime per bag. If you spend $100 a year on French fries, you used to get 200 bags a year when the price was 50 cents a bag. At the new price you get 167 bags: quite a few less fries. And it's not only you, it is every McDonald's customer. For every safety measure the government demands, the price of things served or sold over the counter is bound to go up. All of this happens in order to prevent accidents from occurring in the first place.

The hypothetical decision whether or not the government should order McDonald's to use grease pit covers would come down to a choice of dollars versus risk. The goverment every day analyzes such questions and often decides to order people and businesses to spend money, thus reducing the amount of money available for people to spend according to their own desires. The government argues that it is ordering money to be spent on good things, but the answer comes back that the government is often a very bad decision-maker when it chooses what to spend money on: Look at the boy-in-the-well. The government rarely shops around and chooses the best from among many good things. This is an argument played out again and again. You do not have to sign up irrevocably in the regulation or the anti-regulation camp. Few people do. But extended exposure to different decisions of this sort will eventually develop in you a predisposition in favor of government intervention in the marketplace or against governmental intrusion.

This disposition is the first of your first principles. It is one of the general ideas most discussed in everyday living. People favorably disposed to the free market like to hide the ball of risk. Pro-regulation people like to hide the ball of cost. When you are confronted with a proposal for regulation, ask three questions: Why are we going to regulate? What will it cost if we do regulate? Is there some other way that this money could be better spent?

Being predisposed to favor a free market over a regulated one does not make you a "conservative." Conservatives generally favor keeping marijuana outlawed and banning prostitution and pornography. When first principles collide with one another, one must go overboard. Such choices are easier if you realize that apparent inconsistencies are often only the result of a system of priorities among first principles.

COMMUNISM

There are three ways to experience communism.

The first is to live under an avowedly communist government. If you are reading this book, it is overwhelmingly likely that you are not living under a communist government, because there is not now, never has been, and likely never will be a communist government that allows books critical of communism to circulate freely.

The second way is to study the theory of communism. This is a path many follow, but it is a misleading experience. Communism, no matter which theoretical variety you prefer—Marxist, Leninist, or Maoist—never lives up to its billing. If someone urges you to study the theory without first having studied the historical record of communist regimes, ask yourself why someone so convinced of a recipe's worth is unwilling to allow it to be judged by its results.

Finally, you can experience communism through its history and the written records of those who have lived under it.

Of all the sections of this book most likely to be found

unpalatable at first, this chapter is the best candidate. To think clearly about communism and the big question it poses demands that years of garbage thinking must be cleared away first. Quite a few people resent having their garbage called garbage and will resist the effort to have it removed.

For those of us under thirty it is especially difficult to understand the passions that a discussion of communism unleashes in the older generations. This subject is connected with many controversial figures and many grisly events. Joseph Stalin, Alger Hiss, Richard Nixon, Ho Chi Minh, Joe McCarthy, J. Edgar Hoover—these are just a few of the names tied up in the American understanding of communism. The list of events linked to the subject in the public's mind is long and includes, for example, rampaging mobs, vast prison camps, Vietnam and Korea and their respective wars, and millions of executions. Many Americans guessed wrong about the way that communism would develop and remain embarrassed about their wildly imprecise predictions. A few Americans were traitors, and those that defended their perfidy before it was proved were shamed. Some became resentful. Other Americans were alarmed by these treasons and in their fear, trampled over innocent citizens. It is a touchy subject, so it is best to stick to the facts.

Lenin seized power in Russia in October, 1917, and shortly thereafter set up his secret police. By December, 1919, Cheka—as it was then called—was executing 1,000 men and women every month. By December, 1920, Lenin's 250,000 Chekists had murdered 50,000 people.

Lenin died in 1924. Joseph Stalin succeeded him. Stalin would rule the Soviet Union for three decades. In the first ten years of his rule, at least 10 million Russians died unnatural and unnecessary deaths either as a result of a famine that Stalin engineered or as a result of the harsh conditions of the prison camps he set up. By 1933, the population of those camps had reached 10 million. All of these atrocities were overlooked by observers in the West who were eager to applaud the new, revolutionary experiment in government.

Another tyrant, Adolf Hitler, grabbed power in Germany. The governments of Hitler and Stalin had many more similarities than differences. Today we recognize both men as totalitarians because their ideologies were based on the same objective: total government control of every facet of every individual's life. In August of 1939, Hitler and Stalin concluded a non-aggression pact, and Hitler, freed from a military threat in the East, made total war on the West. In 1941, however, Hitler turned on Stalin and invaded Russia. In the years preceding the war, Stalin had decimated his country's military leadership. As a result, the Soviet Army was woefully unprepared for the double-cross. When peace arrived in 1945, millions more Russians were dead, many as the result of Stalin's trust in Hitler. The labor camps remained full.

Communism swallowed other countries following the war's end. Mao Tse-tung sat atop the communist regime of China. Mass executions, famine, and the anarchy of the Chinese cultural revolution left millions dead.

While these internal upheavals shook China, the Soviet communists and the Chinese communists supported wars in Korea and Indochina. 34,000 Americans died defending South Korea when communist North Korea invaded it in 1950.

More than 50,000 Americans died trying to defend South Vietnam, Laos, and Cambodia from the spread of communism. When America failed, the results were predictable. In Vietnam, hundreds of thousands were herded into concentration camps. Thousands more—the "Boat People"—fled via the sea.

The communist leader of Cambodia from 1975 to 1978—Pol Pot—killed one out of every four of his own people. He and his government were finally overthrown by communist Vietnam's invasion and conquest of his country. Pol Pot—reduced to leading a guerrilla movement against the Vietnamese—announced his "retirement" in 1985. *The Wall Street Journal* paid him the compliment of calling him the perfect communist:

> Pol Pot managed what no other leader anywhere has been able to do: take communism furthest along to its logical conclusion. He emptied the cities, destroyed the currency, pulverized places of worship, drove hospital patients into the fields, ended the use of electricity, wiped out vestiges of education, created a famine, oversaw a resurgence of malaria, and killed one million to three million Cambodians. Far from being mad or uniquely evil, Pol Pot deserves to be remembered as the very model of a modern communist.

Communism has continued to spread. In December of 1979, the Soviets sent 80,000 troops into Afghanistan. Eight years later, the brutalities of that occupation are widely chronicled. They include carpet bombing, mines disguised as children's toys, yellow rain, the massacre of entire villages. We are a little numb to it now.

The story is roughly the same in Cuba. It is now being played out again in Nicaragua. In 1982, for example, the Sandinista government thought nothing of destroying 50 Miskito Indian villages along the Coco River. All the residents were relocated into armed collectives. In 1985, the Sandinistas simply declared an Emergency Decree that ended the civil rights of all Nicaraguans.

These are bold facts. There are a lot of numbers in the preceding paragraphs, but the numbers do not present anything like the whole picture. Many fine accounts of the daily life of the ordinary man or woman living under a communist regime are available. I urge you to read two in particular. One was on the approved reading list in the Soviet Union for a time: Solzhenitsyn's *One Day in the Life of Ivan Denisovich,* a short and interesting novel that is difficult to put down. The other has never and will never circulate freely in the U.S.S.R.: Vladimir Bukovsky's *To Build a Castle: My Life as a Dissenter.* Bukovsky's autobiography should also fascinate you because it chronicles the life of a person who was just about your age when he started a long-running street fight with Soviet authorities. Bukovsky's tale can be unsettling because it asks you what you would have done had you been in his shoes. The young men and women of the Soviet Union have daily thrust upon them a choice: go along with the regime, or resist it. The youth in Hitler's Germany faced a similar choice. The overwhelming number of young people routinely decide to go along to get along. A tiny few find the real courage to resist. This is the courage of accepting repeated beatings, long years in prison, and injections of torturing drugs while wrapped in wet sheets that tighten as they dry.

Books like these can clear your mind. They also immunize you against the silly soporifics of middle and upper class students on the barricades and trendy professors behind the podium.

Even if you intend never to read these books, look back on one I am certain you have been ordered to read: your textbook in American History. That text will no doubt lay out for you in bold relief the American experience. The editions of this decade, however, will spare little ink on communism, its nature, and its history in practice. Two of the three most powerful nations in the world are communist regimes, but you will find in a standard American text very little background information useful to confronting the issues posed by the existence of these regimes and the United States' relationships with them.

How the country's high schools came to a situation wherein both books and teachers sidestep the central issue of our time is a discussion far beyond the scope of this book, though it is a question you might usefully pose to your social studies teacher. He or she might respond that the issue of communism is really the old East-West bugaboo, and the world is no longer so simple that everything boils down to that struggle. Or the response might be that it is more important to understand the United States first, and then to grapple with the rest of the world.

These responses are part and parcel of the new isolationism that flourished in the aftermath of the defeat of the United States in Vietnam. Isolationism is a recurring urge in American history, a collective, national fear of being pulled into wars and disputes not of our own making or not involving our own interests.

This discussion leads up to the question you must answer for yourself:

Should communism be opposed?

Again, this is a deceptively simple phrasing of an enormously complicated riddle. By "opposed," I mean real opposition. Not the soft rhetoric that stresses "co-existence" and "harmony," but real opposition in word and deed.

Real opposition does not mean fighting, though it does entail adequate preparation to fight. Perhaps most importantly it means honesty in assessing that which is to be opposed. When President Reagan named the Soviet Union "the focus of evil" in the world, he was thrashed by opponents of real opposition. His words were deemed a counterproductive reversion to "cold war" rhetoric; but they were honest words.

The stern reaction to the President's explicit description

is rooted in either a lack of exposure to recent history or an unwillingness to confront that history as relevant to the here and now. Recall the earlier question of why the facts in this chapter are not normally found in modern textbooks. If the facts—the simple, undisputed facts—are not available to a broad cross-section of Americans, for instance all high school graduates, we cannot expect that there will be general agreement with the President's statement. If only recent experience, as absorbed through the TV, is the basis for reacting to this cold but clear appraisal of the Soviet Union, then a typical reaction will be: "So the Soviets invaded Afghanistan. So what? We fought a longer war in Vietnam. Our two countries are not that different, are they?"

Of course we are very different. But unless the historical basis for articulating that difference is clearly established in your mind, it will be very hard to avoid semantic traps that subtly minimize the difference between the United States and any country where the ruling government embraces communism.

When you begin college there will be debates over whether our government should support an armed effort to overthrow the Sandinista regime in Nicaragua. That regime is avowedly Marxist-Leninist—a communist government. It came to power through armed revolution and quickly established close ties with the Soviet Union and the communist government in Cuba. Its military is dependent upon supplies from the Soviets. Its legal structure—really, lack of legal structure—and domestic politics are taking on the forms and methods of operation of a standard communist regime.

Many view our government's support of the Contras as immoral. Congress came very close in 1986 to barring aid to them. The debate is a fierce one. You cannot become involved in an intelligent fashion in that debate without first considering the history, the pattern of development, of other communist governments, and answering the question: Should communism be opposed?

Your aim should be to take an objective view of communism, and then to debate and act on the basis of that view. The few facts I gave are a start. The two books I cited are a second step. There are many additional inquiries to be made, important ones. Your college courses and discussions will thrust you into these inquiries. My advice is that you go in prepared with some background and with certain knowledge that none of your partners in the discussion, especially the professors, will fully display the experiences and convictions that color their presentation of the facts.

GOVERNMENT: THE ENDS

This is a short chapter, but do not make the classic error of mistaking brevity for simplicity. Short questions can cloak enormous quandaries. The classic of this category is: Do the ends justify the means? Even to pose the question means posing additional questions. What are "ends" anyway?

"Ends" are goals, targets, aims, purposes. An "end" is where you want to be when the curtain falls, where the plot concludes at the close of a novel. When talk turns to the "ends" of government, the discussion is centered on the condition of the nation that the means of government hopes to produce.

A distinction in terms is necessary at the start of this chapter. Communism is an ideology, a set of ideas. It is not a form of government, although many governments embrace the concept of a communist society, as their goal. Other governments embrace different ideologies and different forms of government because they have different goals, different "ends." The American government shifts around when it comes to ideology from time to time. In basic structure it has remained, since the Declaration of Independence, a democratic republic. Since the ratification of the Constitution, moreover, its goals have been fixed.

The text of the Constitution saves you a lot of hard thinking on what are the ends of the American system of government. The Constitution, our highest law, lays out these goals in matter-of-fact fashion:

1. To form a more perfect Union,
2. Establish justice,
3. Insure domestic tranquility,
4. Provide for the common defense,
5. Promote the general welfare,
6. Secure the blessings of liberty to ourselves and our posterity.

These are the announced goals of our Constitution. The rest of the Constitution provides the means to these ends, a structure of government intended to produce these objectives. A crucial first principle for an American to decide is whether he or she agrees with the British statesman William Gladstone, who said that our Constitution is "the most wonderful work ever struck by the brain and purpose of man." That is, do you agree with Gladstone that the Constitution, and the goals that it sets, are excellent things? We can consider later whether the system of government set up by the Constitution still works, if it ever did, to promote these goals. But for a moment, the focus is on the goals.

You are not wedded to these goals. In fact, the Framers of the Constitution were convinced that if these goals were to be realized, they would have to depend upon a structure that did not demand such implacable allegiance. The "benefits of liberty," for example, cannot be secured by denying one the liberty to heap scorn and throw mud on the Constitution.

So why does the United States have this set of written goals? To give us a measuring stick.

Every high school student in America who has participated in athletics, or even attended a pep rally, has been exposed to "coach talk." Coach talk is a wonderful thing, a universal language that consists mostly of cliches, but which is really the language of goals:

> "Okay, men, this is it. Gut check time. Have you got it in the box? [Coach pounds chest.] You have got to have it in the box or all the talent in the world won't reach our goals."

The goals are very diverse. A 10-0 season. A .300 average. An 18 second split. A six foot jump. A 1:10 quarter. Athletes live and die by goals.

So do governments. The Constitution's list of goals is the coach talk of political theorists. The Framers of the Constitution assembled because the original governing instrument of the newly liberated Colonies, the Articles of Confederation, had failed. The Framers were told to assemble in order to iron out the problems in the Articles. Instead, they promptly exceeded their order merely to tinker, tore up the whole thing, and started over. They produced an entirely new Constitution. The Constitution has fared much better than the Articles, even though it was flawed. So flawed, in fact, that numerous Amendments—the first ten, the Bill of Rights—were needed right away. But except for the Amendments produced by the Civil War, and some other minor additions, the original plan has worn well. One reason the Constitution has endured so well is that the original choice of goals was unusually sound.

41 Government: The Ends

The goals of the Constitution are not mere ornaments. Though they are strung together in the Preamble, and thus invite only a quick glance, each deserves to be separated out, turned over, and examined carefully. The Framers really meant what they said, and the order in which these ends are laid out is itself instructive. Here the Founding Fathers, at the very beginning of our nation as we know it, assembled a concise statement of their priorities.

"To form a more perfect Union": the number one goal. Under the Articles, the states were floundering. Among the problems, according to historian James MacGregor Burns, were "weakness, instability, and inability to control the factious, rambunctious states." The delegates of the Constitutional Convention convened on May 25, 1787, and forging a union—a true, working union of the states—was uppermost in their minds.

"Establish justice": The system of government that emerged was designed to treat individuals with a fair and even hand. Penalties and rewards had to be arrived at equitably and the laws had to be justly administered. The idea of what constituted justice was not spelled out. It would, rather, be allowed to unfold and evolve. Two hundred years ago that idea included the institution of slavery, an institution obviously unjust to us now, and which would be swept away. (Even some of the Framers and Founders were aware of this injustice. Jefferson, himself a slaveowner, remarked that he wept for the country when he considered that God was just, admitting that just men could not abide slavery.)

"Insure domestic tranquility": Simply put, this was the demand for order, as in law and order. Without a minimum amount of order, nothing gets done. Anarchy reigns. People, the citizens, are not secure and are neither content nor happy.

"Provide for the common defense": The old saying goes, "Pray for peace, but keep your powder dry." Having successfully fought a war that was a close-run thing, the Framers realized that without a common defense, all of these goals they were espousing and all the institutions they were building, could be quickly wiped away by a hostile enemy.

"Promote the general welfare": The country would prosper only if it prospered generally, not just for a select few. "General welfare" can be understood as the goal of making money and spreading it around, or more precisely, establishing conditions conducive to all individuals making money. Decades later, an additional understanding was tacked onto this goal, a very noble one: Take care of those who, for whatever reason, cannot make money.

"Secure the blessings of liberty to ourselves and our posterity": The Framers wanted freedom, not only for themselves but for their children and their children's children, etc. "Liberty" meant many different freedoms, including the right to write and read whatever political texts they chose, to pray in whatever fashion they desired, to assemble in whatever numbers and in whatever place they picked, and to simply bolt the country if they decided to do so. The Framers' attachment to liberty in all these and other forms is a defining trait of that group.

Hosts of books have been written about these goals and their meaning. Many arguments spin on and on concerning their relevance today. Many people challenge their usefulness. One easy way for you to consider their merit is to think about those goals that are not included in the Constitution's Preamble. The Framers did not, for example, say:

"Extend the dominion of the government over additional lands": The Framers were not land-grabbers, were not necessarily acquisitive imperialists who conceived that the nation's borders would grow and grow. But neither did these men rule out the possibility of expansion. Additions of territory were to be considered as the proposals came along.

"Insure the equal condition of all citizens": The Framers were men with money, and were well aware of the different ways men could come by money—inheritance, work, theft. They did not mandate that everyone should have the same or even roughly the same amount. They left open the question of how wealth ought to be distributed.

"Promote the physical health of the citizenry": There is no Constitutional fitness program, no national health plan. Sloth is okay. So is drinking and smoking.

"Exterminate the Indian Tribes": Eventually, the national government would follow policies that would do just about that. But when the actions were taken that contributed to this result, they were perceived as necessary means to Constitutional ends. In retrospect, we see clearly that these actions were not just. But that injustice was not enshrined in the Constitution.

The list of goals not included is endless. That infinite list of possibilities should impress on you just how careful a choice the Framers made when they selected the language for the Preamble. They could not aim for everything, but they did aim for what they believed to be the most important things.

When college surrounds you, there will be much debate on these goals. Many are convinced that the country needs to add new goals in the form of new Amendments, such as a complete ban on abortion or a guarantee of equality for women. Others argue that no new tinkering is necessary.

Among your first principles should be a judgment of whether the Constitution's goals are good ones, bad ones, a mixed bag, or simply incomplete. Because the Constitution's goals are the goals of the nation, and you are a citizen of the nation, they require of you an affirmation or a rejection. This internal discussion and decision must precede any external debate you have with others about the particular policies—the means—of government. These external wrangles can often be confused because the parties to them have not got clear in their own heads what all this government stuff is supposed to be about. Once a set of goals—ends—for the government is established or advocated, the particulars of any given policy become much easier to judge because one key question is always obvious: Does this particular policy contribute to the realization of the goals of the government?

The choice of goals is not easy. Very worthy goals can compete with each other, can even come into inescapable conflict. When that happens, tough choices loom.

Intelligent choice demands that you establish a priority, because government cannot do all things. Those things that are at the top of the list win the appropriate recognition. Indeed, in this country, they are written into the highest law, the Constitution, and policies contrary to their achievement are not permitted.

Among your first principles should be an evaluation of the country's current goals. This chapter is just a windy way of pointing to that list in the Preamble and asking: Do you agree? Are the Constitution's goals the ones the country ought to be pursuing?

GOVERNMENT: THE MEANS

There are six basic forms of government: monarchy, tyrany, aristocracy, oligarchy, democracy, and anarchy. All other labels are variations on these forms. These six divide easily into mirrored pairs: monarchy and its corrupt version, tyranny; aristocracy and its corrupt form, oligarchy; democracy and its corrupt form, anarchy.

The American experience is with one of the many types of democracy: republican, representative democracy. The "republican" part comes from the participation of fifty smaller governments, the states, in the one national government, the federal government. The "representative" part comes from the fact that Americans generally do not vote directly on issues, but instead elect representatives to conduct the business of government for them.

Student government in America is a marriage of democracy and aristocracy. To be sure, the forms are all democratic, as the student councils of the nation's high schools are all under the rule of the ballot box. But as any realist knows, a student government depends on the good will of the student elite: the athletes, the cheerleaders, etc. These groups make up the school's aristocracy.

Student government has two obvious purposes: to pad the resumes of its participants and to deal with an agenda of minor things over which the students have control, like dances, class rings, pep rallies, and class reunions in the future.

Lost in the mists of education history is a third purpose of student government: It is a vocational course in the means—the instruments and institutions—of American government.

Student government would go a long way toward improving its usefulness if it shed these democratic forms. Why not, for example, a student government based on monarchy? Pick a student king, give him absolute control of the student agenda and the student treasury, and allow him (or the monarch Queen) to appoint the successor. More often than not he or she would rule like a petty tyrant, but the stakes and cost would be small and the education large.

Or the informal aristocracy could be made legitimate and power ceded to it. Football lettermen, even the linemen, could rule by consensus or an agreed upon division of authority. You guess the results.

Good ideas, both, but inevitably suppressed and sacrificed by the American homage to democracy. Too bad, because even a minimal exposure to the alternatives to democracy would prompt a good many students to think carefully about the American means of government.

The preceding chapter listed and briefly described the avowed goals of the Constitution. The most important vehicle by which those goals were to be pursued and attained was also written into the Constitution: a limited, central, federal government divided into three branches. Each branch was given a unique role.

You should know this already, and know also of the growth of power of the Fourth Estate, the media, and the

Fifth, political action committees. If you don't know these things you have been shortchanged in high school, and you are dead meat in college. You will be the source of much sport among your savvy acquaintances. Dust off your government text. Even if you are a science whiz-kid, you will have to figure out how the government works if only to know where the grant money comes from.

The Framers knew that the central, federal government had to have power if it was going to get anything done. The Articles of Confederation had failed primarily because that grant of power was missing. But they also feared, indeed feared foremost, the concentration of power in too few hands. George III left a bad taste in their mouths.

So they contrived to spread the power around. A lot of people would have to agree if the government was going to do something. The two halves of Congress, the House and Senate, would have to agree on a proposition. Then the President would have to sign on. If he didn't, the proposal was vetoed and even greater numbers of Congressmen would have to agree to get around the veto. Even if Congress and the President agreed, and a proposal became law, the Supreme Court would eventually have the opportunity to declare: "It is unconstitutional. Start over." Nowadays, special interest groups can take aim at a target at any point in this elaborate process and pull out the long knives. And the media perches on the sidelines calling out its view and illuminating/intruding upon the entire process.

It is a circus. And it works well, or at least less badly than all other plans.

Do you approve of the division of power into so many hands? It is a basic question.

Many people fall into the error of judging the American means of governing on the basis of particular results on given issues. Upset, for example, with the tax code's treatment of foreign income or the Supreme Court's abortion decisions, they reject the entire system. But reactions to particular issues are rarely first principles. It is possible to, as a first principle, admire and love the system, but despise any number of results.

Do you like the system?

The only significant change in the system's basic design since the Framers signed-on has been the exponential growth in the power of the courts. The press has always been powerful. But the power of the courts to control the details of American government and American life has soared.

Courts have assumed a position vis-a-vis the rest of government similar to that of school authorities vis-a-vis student governments. At any time, for nearly any reason, the courts can overrule the will of the rest of the government, and the school principal can toss the student council's plans in the trash.

Intervention by courts and school administrators exasperates elected representatives in both forums for many reasons. The most important reason is difficult to articulate. It boils down to the losers' inability to fight back. The students can do nothing to strike back at the school authorities, and Congress and the President have next to nothing in their arsenal that is effective against the Supreme Court. Principals and judges are thus, to a certain extent, above the law, because they announce the law, and disagreement is not possible. A good or bad thing?

You cannot answer that unless you have first thought through your views on one of the first principles of American government: The people who make up the government ought to be accountable to the people it represents.

Accountability is a large word but a simple, and powerful, idea. It is the right to throw the bum out.

Student governments are accountable to some degree,

but not very. If the smoothie whom you elect as a class leader turns out to be a pushover with the school authorities, or dumb, or hopelessly behind the pop culture curve, you can throw him out at the next election. When your Congressman, Senator, or President is a Bozo, you can do the same thing.

But if the clown is occupying the principal's office, the students can do nothing. If he or she has donned judicial robes, the voters will never have a direct opportunity to fire that judge. Period. The principal is completely unaccountable to the students, and the judges of the federal government are completely unaccountable to the American voter.

Do you like the idea of the accountability of government officials? Or do you prefer to give grants of immunity from public opinion? One such grant is lifetime job security. Judges have that security.

When you consider the means to the ends of American government, the two primary means are up front.

First, power is split up among a lot of people. Do you approve of a system that divides authority among a few hundred hands? This division makes it difficult for the government to act, but it also protects you from the situation of too few people running the show and doing so in unpleasant fashion.

Second, most government officials are accountable. Would you prefer more, less, or the same amount of accountability as the current mix presents?

The preference for power-sharing and for accountable officials is evident in numerous American institutions. These preferences can be personal as well as institutional.

When you decide to seek a job or join a church, for example, the question of the division of power within the group and the accountability of the boss or pastor may figure in your decision. Does the business run on the final say of a single executive, or is authority widely delegated? Does the pastor seek the advice and guidance of his entire congregation, or does he handle church affairs entirely by himself? For some, these inquiries and the answers they elicit will not matter. For others, they will be decisive.

I am suspicious of too much power in one place. It is sometimes a necessity, though. Foreign policy cannot be run by committee, and the decision to go ahead with the D-Day invasion of Normandy, when the weather conditions were stormy and the chance of their clearing iffy, could only have rested on one man. When General Eisenhower turned to his advisors and said, "Let's go," he assumed responsibility. On many issues, decisions must be put to, and taken by, one individual, or no decision will be reached. But I think these decisions are few and far between.

I am also suspicious, very suspicious, of people with power over my life who are not accountable to me, or who are accountable to me only when I organize enough similarly situated people to act. I worry, for this reason, when I enter into a dispute with a large organization, because that organization's bureaucrats have no incentive to take my complaint or my request seriously. If a bureaucrat acts expeditiously, I am surprised. There is no reason other than personal honor for him to do so.

Not all people, even all government officials, can be made accountable. Regulatory bodies, like the Food and Drug Administration, must be made immune from outside pressure if we want to be confident of the impartiality of their decisions. So too must judges, to some extent, be immune from meddling if we want them to act fairly and impartially and not with an eye on their re-election campaign.

Great power in the hands of a single person who is not at all accountable, or only remotely accountable, poses the great danger of power abused. And yet this was exactly Eisenhower's position on the eve of D-Day. Obviously, we cannot rule out the necessity for such a situation occurring again. The ability to judge the wisdom of future concentrations of power rests in an understanding of the Constitution's design. Only when the norm is grasped can the abnormal be evaluated.

RACE

It is difficult to discuss race without outraging someone. Race figures in American history like a body-length scar, and it is not yet fully healed. It is an ugly thing to look at, and painful to touch. But it cannot be avoided.

The University of Michigan Law School is not a particularly race-conscious place. It is probably in the mainstream of American campuses. There was, when I was there, a black students organization, an Hispanic students association, and a Jewish students organization. There will be similar organizations at your undergraduate school. Membership, by definition, is not open to all comers.

At U. of M., and at Harvard, there were predominantly white organizations as well. By predominantly white I do not mean those organizations that simply reflect the distribution of races in the student body. I mean organizations that are conscious of how many minority members are counted on their rolls. The Final Clubs—upper-crust fraternities—at Harvard are a prime example.

Campus organizations that are run without regard to

race are few and far between. That is a tragedy, but also a fact. Those that are neutral on the question, such as organizations devoted to the arts and sports, are pure meritocracies. Either you are the best soloist on bass or the best goalie on the ice, or you are not. Race does not figure in. Social groups, fraternities and sororities for example, enroll members on other grounds, and racial decisions seep into the process.

The central question posed by this chapter is: Ought race to figure in any decision that bestows a benefit or inflicts a penalty? It is a peculiarly American question because our country is home to people of many different races.

Not so long ago, the Ivy League colleges would accept for admission only a certain number of Jews. Today, some of the same colleges and their elite counterparts in other regions have been accused of limiting the number of Americans of Asian descent who are admitted into the colleges in any given year. The same colleges are vigorously recruiting black and Hispanic students. Race figures heavily in getting in the door. It should not surprise you that it continues as an issue once the successful applicants gather and begin classes.

Colleges deal with these issues gently. Outside of academia the problem of race is not susceptible to such soft handling. Most discussions flame with passion. Many believe that the well is poisoned on this subject and that no calm discussion is possible, or even should be.

Successive generations of Americans, from the Framers forward, have been guilty of various kinds of severe racial injustice. Americans first applauded and then tolerated slavery. The Civil War was fought, not in the first instance to abolish slavery, but to save the Union. The Emancipation Proclamation that ended slavery was a by-product of Lincoln's quest to save the Union. For a century afterwards, Americans applauded or tolerated segration in many areas of the country. American history also includes the systematic destruction of the Indian tribes, the internment of Japanese-Americans in prison camps during the Second World War, the failure of the American government to come to the aid of Europe's Jews as Hitler methodically murdered them, and many other shameful episodes. Today we are trying desperately to close our southern border to Mexicans who come here to find work, and to fend off Asians who have fled the various tyrannies in the East. Those who somehow get in often find overt hostility in the towns and cities in which they settle. The Cubans who fled Castro in the great boat-lift have inspired not a welcome but hatred from many people in Florida. Arabs are routinely hooted at. An American of Asian descent was beaten to death in Detroit a few years back by a couple of Americans who were allegedly outraged by Japanese competition.

There is a lot of hate in America. A great deal of it springs from a consciousness of race.

The decision to bus children to integrate the schools, the decision to admit only so many minorities to an organization, the decision to award certain jobs only to minority applicants, the decision to give contracts only to minority-owned firms—all of these decisions presuppose an initial decision to take race into account in giving out a benefit or inflicting a penalty. Is that initial presupposition a good one?

I had a law professor at Michigan who used to press hard on this question, though in a roundabout way. Look, he'd say, assume you are going to use race in making decisions like these. How are you going to measure race? Then he would start a string of hypotheticals. Assume a black American marries a Norwegian-American. Does their child qualify for "minority" status? If so, he'd continue, assume that child grows up and marries a German-American. Does the child of that second generation marriage qualify as a "minority"? Whatever your response to either question, he would then ask why you took that position.

The system for the award of benefits due to minority status now depends on self-selection. Recall your college applications, if you have filled them out already. Somewhere on each of them was a box which you could fill in at your option. The box asked you to identify your ethnic heritage. You filled it in. That is an example of self-selection.

The colleges publish no rules on what constitutes a credential as a minority. My professor liked to ask whether such rules would have to be developed, and if so, by whom? Millions of dollars of benefits are already at stake, he'd explain, and the amounts will only go higher. Are such rules necessary if only to distinguish between the competitors for minority benefits?

The alternative to race-conscious distribution of benefits is race-neutral distribution. No boxes to be filled in, period. Awards and penalties would depend on other criteria. There are three arguments against this approach.

The first is that there is a chance that minorities will be discriminated against in the award process and that preferential treatment guards against and corrects such discrimination. Unless benefits are directed to minorities, the argument runs, they will never flow in that direction naturally. Opponents of race-conscious selection argue that actual, intentional discrimination on racial grounds is illegal, and that people who do discriminate can be caught and punished.

The second argument is that minorities labor under mighty disadvantages. When it comes time to apply to college, how can a kid from Harlem who struggled through an inner-city school, often without the benefit of two parents or even two decent meals a day, be expected to compete on an even basis with the upper-class kid who went to a private prep school and summered in Europe? How can a minority businessman running a small operation be expected to compete against corporate behemoths like General Motors? Opponents reply that these are false dilemmas, and that the minorities who benefit from race-conscious award processes are frequently themselves the products of upper-class upbringing or large-size corporate settings. But this reply sidesteps the argument on its merits. Unless every beneficiary of a race-conscious award is in fact upper-class and competing on an equal basis, then the reply is insufficient. So too, however, is the argument because it does not consider that a poor white kid or a poor white businessman may be edged out from the opportunity to advance solely because his or her competitor is a minority.

The third argument is the demand made by history, and it is particularly relevant to the situation of black Americans. This argument incorporates by reference not only slavery and segregation, but also the lengthy record of overt discrimination and hostility towards blacks that continued even after the Civil Rights Act of 1964 and the Voting Rights Act of 1965. "We were wronged as an entire race," the argument runs. "Every bigot hated every black, and every act of bigotry was aimed at all blacks. Individuals were not judged as individuals, but an entire race was enslaved, an entire race segregated, and an entire race wronged. So all blacks deserve restitution." Restitution is at the heart of this demand: Make whole that which was injured. Restitution is a familiar concept in the law of injuries. If a man loses his arm as a result of your actions, you pay him for that loss. This restitution argument in the racial setting demands a just compensation for all the wrongs done to black Americans, all Native Americans, all Hispanic Americans.

But does restitution have any place in the context of injuries done to an entire group? What about a young, upper-class black youth who, it turns out, is a gifted scholar? Should he receive restitution in the same proportion as the Harlem kid? And those who pay, even if only in terms of opportunities forfeited to affirmative action programs, is it just to charge them for injustices in which they took no part?

There is also the question, "How long?" Even if we accept the idea of benefits awarded on the basis of race, for how many years are these race-conscious awards going to continue? Five more years, or ten, or thirty? If there is a legitimate demand from history, when will it be satisfied?

I am bothered not only by this debate, but also by the great unspoken debate. How does this disagreement connect up with the problem of racial hatred? Entitlement and restitution aside, what about the underlying problem of race-consciousness? The "white-dominated" institutions on your campus or in your community are going to be there regardless of whether benefits are bestowed or penalties inflicted on the basis of race. The issue of ending race-consciousness and thus ending racial hatred, is not on the public agenda.

It is absent not because the issue is unimportant, but because it is hard to think of what new, collective steps we can take to deal with it. The government long ago passed laws outlawing racial discrimination in housing, hotel and travel accommodations, and voting procedures. But the government cannot outlaw racist attitudes. You cannot send people to jail or fine them because of anti-Semitic jokes or other racial slurs. Those who utter such slurs can only be ostracized, but even that is infrequent. When was the last time you heard someone say to the racist joke-teller: "That's offensive, you jerk." It does not happen. Even right-minded people do not challenge the "Heh, it's only a joke, I'm not serious" bigot.

Because you cannot outlaw racism and racial hatred, it will remain with us until race-consciousness doesn't exist. And I doubt that race-consciousness will ever disappear so long as race is an issue, a deciding factor, in dividing up benefits. When race no longer figures in issues of public policy, it will begin to wither in individuals' minds, but not until then. The unspoken debate is whether this is true, and if it is, whether the end of race-consciousness is a good thing and something to be aimed for.

You need not agree with my conclusion. Even if you do, you may also conclude that restitution is more important in the short-run than ending race-consciousness and thus racism. But the need is immediate for you to consider the issue of whether race ought to play any role in deciding who gets what slice of the pie. If you have applied to college and have seen that box, you already know that the issue is out there. And it has already affected you.

FRIENDSHIP

This chapter is the circle forced to fit the square.

You have gone halfway through this primer, and some of the chapters thus far have concerned fairly dense stuff. This is the break, a time-out of sorts. It is largely unrelated to all but two of the other subjects. Unrelated in intellectual terms, but friendship inevitably affects how we come to various ideas and the decision to keep or modify them.

You may have noticed a great meanness in American public life. Political opponents, especially those slugging it out in campaigns for elected office, are capable of doing and saying tremendously underhanded things, and many appear to enjoy it. Campaigns only make obvious what is a constant element in politics—in Washington especially. Many people in pursuit of their agendas do not hesitate to slander their opponents or even non-combatants who wander into the line of fire.

This is definitely not a new development. Scoundrels have been around since colonial days. But today the march of mean-spiritedness is greatly magnified and its effects more widely felt as a result of the increasing power of the media. Folks who willingly blacken eyes can do so in front of hundreds of thousands of viewers. Reporters appear to revel in doing harm to reputations. They are eager partners in taking any venomous message public.

The public eats it up. For one thing, there is great drama in name-calling and scandal. Nothing is more likely to lead-off the evening news than a shouting match between political opponents because it is such good theater. Ugly charges are thrown around and TV picks up and replays them because they entertain the viewers. So does scandal.

But TV is not to blame. TV and the media generally are the pack mules of today's political environment. They'll carry anything, anywhere without so much as a single kick if they are well fed at the end. If the ratings show the public wants it, the public gets it.

Meanness is a part of the public culture because meanness pays off. Rolling someone is an effective way to get rid of opposition. Hyper-criticism that calls into doubt an opponent's intelligence, character, patriotism, or motives can demolish an opponent, and his or her ideas go down with him.

The resort to mean-spirited tactics is not the preserve of the Right or the Left. At the extreme Right or the extreme Left pure cruelty is often apparent, but even among the merely "right-wing" or "left-wing" it is obvious every day. A by-product of this development is that the political description "moderate" almost always attaches itself to pleasant people.

In recent years, the Right has been more visible and, as a result, its various cruelties have been easier to spot. Mocking sarcasm is a favorite weapon of the Right. But it is used by the Left as well.

One of the most visible examples of acceptable meanness is Gary Trudeau's "Doonesbury" cartoon strip. That strip so often abandons civility in pursuit of black humor that it no longer surprises. Of more limited circulation but equally talented when it comes to the tactics of derision and bullying is conservative columnist John Lofton. If you have never heard of him, you are lucky. His style is one of noisy confrontation and frequent explosions of wrath in print. His ideas are often reasonable, but his approach is so bare-knuckled that I've lost my capacity to read him anymore.

You may already be a viewer of Saturday Night Live. If not, and if things have not completely changed since 1978, my last year as an undergraduate, you will watch this show on a fairly regular basis. It is the perfect example of "humor" used to disguise baseness. Fully half of the skits during the years I watched were meant to mock, and injure, and embarrass. The defense can be put forward that satire has always been around. My response is, yes, that's true, but now it has hit new lows.

There is nothing that will change this situation. Meg Greenfield, editor of the *The Washington Post* editorial page, wrote a column in 1983 in which she made the perfectly true point that almost everyone in Washington makes jokes about the city's government which is run primarily by black officials. These jokes are tinged with racism, she noted, but remain acceptable in private conversation. "One is treated to endless snickers about the incompetence in our city government," she observed. She decried the situation and denounced the hypocrisy involved, but it probably hasn't changed one bit.

High school students are very accustomed to cruel, slicing humor. The class geek is no doubt a frequent target, or the gawky young woman, or the unfortunately overweight of either sex. You do not have to be convinced of this point, only reminded.

At college, the cruel edge to humor sharpens as it picks up sophistication and vocabulary. When I was in law school I bused tables at an undergraduate sorority. The claws that came out at dinner were long indeed, and I was relieved that the ladies were never angry with the help. When the end of the year rolled around and the hired hands were given a dinner, we in turn presented a pitcher of ice to one of our employers, naming her publicly the ice-queen. Cruel humor, that. It got quite a laugh from the women, the great majority of whom appeared to share our judgment. The recipient was visibly embarrassed.

As that example shows, humor has become the blanket pardon covering and forgiving mean-spiritedness, and mean-spiritedness continues to proliferate. Such is the way of the world. What has this got to do with friendship?

It is connected to friendship because friendship is the quality most studiously ignored by the people who put together the American government. Those people, the Framers, counted on cruelty and mean-spiritedness as a defining quality of the human condition.

When James Madison set out to explain to the states the necessity of their ratifying the new Constitution—in the tenth paper of the Federalist series—he did so with his eyes open to the basic problem of government, which is the cruel tendency of those in power to oppress those who are not in power:

> The latent causes of faction are thus sown in the nature of man; and we see them everywhere brought into different degrees of activity, according to the different circumstances of civil society. A zeal for different opinions concerning religion, concerning government, and many other points, as well as speculation as of practice; an attachment to different leaders ambitiously contending for pre-eminence and power; or to persons of other descriptions whose fortunes have been interesting to the human passions, have, in turn, divided mankind into parties, inflamed them with mutual animosity, and rendered them much more disposed to vex and oppress each other

> than to cooperate for their common good. So strong is this propensity of mankind to fall into mutual animosities that where no substantial occasion presents itself the most frivolous and fanciful distinctions have been sufficient to kindle their unfriendly passions and excite their most violent conflicts.

Stop. Read that again.

What does Madison tell you? Among other things, he tells you that not much has changed in the past 200 years. Men and women still love to fall to fighting and bashing one another. Denied big, grand reasons for doing so, "frivolous and fanciful distinctions" will serve. Madison's language is elegant, but the situation he describes is squalid.

The Constitution Madison argued for was meant to address this reality of everyday life. No bromides about utopia surfaced in the Federalist Papers. "If men were angels," Madison wrote in Federalist 51, "no government would be necessary." But of course men are not angels, so the Framers set about divvying up power and providing protections for small groups and individuals. They did not attempt to cure the very human evils of violence and cruelty. Rather, they gave them full freedom to play themselves out in print. This is only one benefit of the First Amendment, but it is a large and significant one: You can say just about anything concerning a public figure and get away with it. Cruelty and meanness were given and continue to enjoy an outlet. They have taken full advantage of it.

So little minds blow, blow, and blow hot air, racist jokes, slanders, lies, and all sorts of calumny. Names and reputations are routinely trashed and vicious humor is a staple of our lives. But only two groups suffer. The victim, of course, is sacrificed to the more important end of providing the mean-spirited a relatively harmless outlet. And the blowhard himself, who is diminished and hardened by his continual resort to smears and jealous comment. But the bitterness that is vented this way rarely hardens into physical harm.

The craftsmen of the Constitution did not give much thought to saving the character or the souls of individuals. They set out to save the country, and they pulled it off. Others took up the first task, including churches and, more recently, self-help hucksters.

Philosophers have also tried to lend some aid in this area. There is an audience for these counsels. Who, after all, wants to be remembered for callousness, cynicism, and a venomous pen or tongue? Oh, there are some. But generally mean-spiritedness is not so much a reputation one strives for, as a habit into which one falls. The condition of public disagreement on public issues is so frequently poisoned that this in turn contributes to an easy descent into incivility in private life. The public can be expected to mirror the traits of celebrities. How to avoid the fall?

The title of the chapter tips my hand, but it also leaves a great deal unrevealed. The varieties of friendship are numerous. Friendship is an accordian word, capable of stretching or contracting to include or exclude a variety of conditions. The common sense reflections that follow have in mind a particular kind of friendship that includes many standard ideas and one twist.

You are no doubt already acquainted with friendship. You have pals, buddies, comrades. You enjoy their company because you have like tastes or experiences that somehow brought you together and a mutual admiration that keeps you together. Strong friendships are marked by affection and candor. Loyalty and trust are also their traits, and, typically, good humor.

The mutuality of tastes is not necessarily extreme among friends. A single, common interest can start two people on the road to friendship. They may share nothing else in common. Good friends can revel in their differences.

Those differences can include absolutely irreconcilable disagreements. One friend may be a resolute atheist, the other a fervent fundamentalist Christian. A dedicated communist can be fast friends with a libertarian. A passionate Dallas Cowboy fan can befriend a Redskins fanatic. "I never considered a difference of opinion in politics, in religion, in philosophy, as a cause for withdrawing from a friendship," wrote Thomas Jefferson. Jefferson's friendship with John Adams, forged during the Revolutionary War, did falter for a time. The two men were political adversaries of the hottest kind: They both stood on principles they devoutly believed to be true but which the other was confident were false. Eventually, at the urging of intermediaries, they resumed a correspondence and resurrected a friendship, though both remained intensely partisan. These two great antagonists and friends died on the same day, July 4, 1826. The friendly rivalry continued to the end. On the day they died, Adams awoke from a coma and spoke only the words: "Thomas Jefferson survives." In fact, Jefferson passed on a few hours before Adams.

This country's Civil War proved how far friendship could be stretched yet still survive in the face of extreme political disagreement. At Gettysburg, the Union General Winfield Scott Hancock and the Confederate General Lewis Armistead faced each other from opposite sides of the famous Pickett's Charge. Their friendship survived though Armistead did not. Bruce Caton recounted the story.

> Back in the spring of 1861, when the country was just breaking apart and officers of the regular army were choosing up their sides, there was a farewell party one evening in the officers' quarters of the army post at the little California town of Los Angeles. The host was Captain Winfield Scott Hancock, and he was giving the party to say good-by to certain Southern officers who were

> going east to Richmond, where they would take commissions with the new Confederacy. The departing guests were sad—it was not easy for those regulars to cut loose from the army they had given their lives to—and a tragic shadow lay across the little gathering . . . . Major Armistead came over and put his hands on Hancock's shoulders, tears streaming down his cheeks, and said: "Hancock, good-by—you can never know what this has cost me." Then the guests left, and next morning Armistead and the others started east, and a little later Hancock himself came east to fight on the Northern side, and he and Armistead had not seen each other since. Now Hancock was on his horse on Cemetery Ridge, waiting with the guns all around him, and Armistead was coming up the slope with his black felt hat on the end of this sword, and the strange roads of war the two old friends had followed were coming together at last.

Pickett's Charge—"for the last time in this war, perhaps for the last time anywhere, the grand pageantry and color of war in the old style, beautiful, and majestic and terrible: fighting men lined up for a mile and a half from flank to flank, slashed red flags overhead, soldiers marching elbow to elbow, officers with drawn swords, sunlight gleaming from thousands of musket barrels, lines dressed as if for parade"—was repulsed, the last gasp in the battle that marked the high water mark of the Confederacy, a battle that would kill or wound close to 50,000 men in three days. Both generals were wounded during the charge:

> Hancock was on a stretcher, dictating a note to Meade. He believed that a quick counterattack now would take the Rebel army off balance and finish it, and he urged that the men be sent forward without delay. He added proudly: "I did not leave the field so long as a Rebel was to be seen upright." An aide came up to him and handed

> him a watch, a pair of spurs, and other trinkets. They came from Lewis Armistead, whom the aide had found dying. . . . Armistead had asked that these mementos of an old friendship be sent to Hancock, and he had gasped out some sort of a farewell message. "Tell Hancock I have done him and my country a great injustice which I will never cease to regret," was the way the aide had it; he may have dressed it up a good deal or, for that matter, he may have dreamed it all, and it does not matter much either way. Armistead had died, going beyond regrets forever, and as if he had been waiting for this last message, Hancock had the stretcher bearers carry him off to the field hospital.

The story illustrates that friendship is quite incapable of preventing even the unusual situation of good friend seeking to destroy good friend. The writer E. M. Forster stated that if it came down to the choice of betraying his country or betraying his friend, that he hoped he would have the guts to betray his country. That is the simple, theoretical conclusion of someone not faced with the choice. As the Hancock-Armistead vignette illustrates, the power of friendship cannot always work to avoid confrontation or reconcile great differences in convictions. But that does not mean that great differences preclude friendship. The passions that ideas and convictions ignite can overwhelm the normal course of a friendship. But friendship can act as a check ensuring that passion does not overwhelm civility.

There is a good argument to be made that diversity in tastes and convictions provides the best environment for a friendship, because these differences inevitably affect and change the two friends. Perhaps the opposing tastes will not have a great impact on each other, but they do at least provoke an attempt at appreciation.

Friendships that cross political lines are also a great

defense against mean-spiritedness, and an antidote to ruthlessness that has already taken root. Those who are friends and also close to each other in all political opinions feed on each other's passion. They agree and agree and agree. Their certainty can become rigid and evolve into blind arrogance. They lose the ability and the willingness to concede anything to opposing views, particularly good faith. The result is the modern ideologue, a term that understandably does not convey much of an image of compassion.

Friends in opposite political camps are apt to yell and shout at one another in private. To thrust and parry and criticize and bitch and moan and groan. Passionate differences can spur arguments that run for years. But there is little bitterness and no animosity in these disputes. Those two qualities cannot co-exist with healthy friendship.

What can happen to friends of decidedly different views? Perhaps without noticing, they modify their approaches to an issue. They account for their friends' views. They consider these views, though in the end they may dismiss them. But the friendship strengthens rather than diminishes their capacity for mutual tolerance. Tolerance is the defense against the awful cruelty of an unchecked ideology.

The word respect is much bruised by overuse. So many peddlers of cheap psychology use it to describe a positive attitude that it has lost a great deal of meaning. It has become trite to urge respect for opposing views. I think the term ought to mean the ability to see and understand the reason in any argument, even if it is a lousy argument, and to say quietly that there is room for doubt. Respect is, at base, the ability to see in one's self the capacity for great, even awesome error: "I could be seriously wrong about all of this." Friends inspire respect in their friends.

Can friendship cure the human tendencies that Madison described, and titled the evils of faction? No. Can it stop war, greed, or famine? No. It can, however, temper passion, promote compassion, and produce better thinking. Even when friendship fails to prevent tragedy, and tragedy results, it does one salutary thing. It induces regret, a valuable if much overlooked emotion that has done more to prevent evil than we can ever hope to know.

So what's the point? Be open to, indeed search out the unlikely friendship. When my roommates graduated from college, one went to work for Father Robert Drinan, another for Tip O'Neill, another finished his work for Senator John Glenn, and I went off to California to work for Eisenhowers and Nixons. We were a strange group of friends for people who figure friendship as at least a partial function of political belief. One fellow couldn't recognize a communist if Lenin's embalmed body rose up and walked into his room. My former roommates are certain that I have never known real compassion for the truly poor in America and that I never saw a federal program save defense that I wouldn't love to throw away. But one of my friends is aware that others think him absolutely blind to the threat of communism and I am aware that others consider my attitude to the poor to be as suspect as any other rich Republican's. We both feel that these are unfair charges. The perceptions of others have proven to be powerfully moderating influences on our zeal, particularly, perhaps only, because those perceptions are held by good friends.

"There is a magic," wrote Disraeli in *Endymion*, "in the memory of a schoolboy friendship; it softens the heart, and even affects the nervous system of those who have no heart." College offers the opportunity to strike up the friendships that will provide a storehouse full of that magic. For all the reasons listed above—the avoidance of mean-spiritedness, the development of compassion and perspective, and the evolution of the capacity for regret—do not blow this opportunity. Avoid the temptation to surround yourself with mirrors.

DEFENSE

How tempting to title this chapter "Nuclear Weapons" and just run with the sexy stuff. God knows this is the preferred tactic of most people who write for general audiences in this area.

My favorite defense experts do that, as do "experts" I consider to be loony. The problem with that approach is that it presumes too much: It presumes the audience has the basics down. Let me illustrate the difficulties that can arise when a speaker or an author simply assumes that the audience knows the background. Richard Perle is an Assistant Secretary of Defense and a very controversial figure in the never-ending debate about defense policy. In 1985 he spoke to a conference about the problem of nuclear proliferation, and *The Wall Street Journal* excerpted and reprinted a portion of his remarks:

> Ladies and gentlemen, we in this room and most of the world accept an image of the strategic relationship between the United States and the Soviet Union that is characterized by a spiraling arms race. And yet the facts

> are significantly different. The United States has today, deployed around the world, some 8,000 fewer nuclear weapons than we had deployed in 1967. And as Senator Stevens indicated earlier, the megatonnage of this diminished American force is barely one-quarter of what it was in the late 1960's. Moreover, the western alliance agreed, at a meeting in Canada a little over a year ago, to reduce further, by 1,400 weapons, the number of our nuclear weapons in Europe. By contrast, we have seen in recent years consistent additions to Soviet nuclear forces: 8,000 new strategic warheads alone since 1969, when the SALT I negotions got under way, 4,000 of which have been added since 1979 when the SALT II treaty was signed.

What is Perle's point? He appears to reject the idea that the U.S. is engaged in a nuclear arms race. He seems to believe, rather, that the Soviet Union is piling up nuclear weapons while the U.S. is cutting back its stockpile. Harold Brown, Secretary of Defense under President Carter, made the same point more succinctly: "When we build, they build. When we stop building, they build."

But what is the premise behind Perle's point? Who knows? This is only an excerpt from what was probably a long speech. It is more interesting to ask: What was the premise of *The Wall Street Journal* in printing this excerpt? The editors must have concluded that the few facts in this paragraph are very important, that these facts will somehow instruct the paper's readers. The premise seems to me wrong, because facts like those used by Perle and reprinted in the *Journal* have very little impact on a reader who has not thought through the first principles of defense. They are simply another set of numbers floating around. They are part of a complex corollary to an even more complex logic of defense theory that is rarely, if ever, broken down into basic components.

While some of the participants in the debate over defense theory are tangled up in detailed analyses, number comparisons, and computer progressions, others shrink from even a minimum level of complexity, a level that serious talk about defense requires. My favorite example of this retreat from reason is the bumper-sticker: "You can't hug a child with nuclear arms." Another bumper-sticker reads: "One nuclear bomb can ruin your whole day." There are others. All of them are flat-out stupid. They lack even the ironic humor of my candidate for sticker-of-the-year—the brown background ribbon with a Smokey-the-Bear profile and the words: "Support Your Right to Arm Bears."

You do not have to retreat from the complexity of Perle's speech to the idiot appeal of bumper-stickers. It is possible to master the basics of defense theory and build from there. The trick is going about this process in the right way.

Recall your primary school and junior high years. Those eight or nine years included an almost daily lesson in a subject called "Math." Math class included a lot of different topics, from addition, subtraction, and division to fractions and decimals. When you hit high school, "Math" went off the course selection list and you were offered a progression of classes that typically runs: Algebra, Geometry, Trigonometry, Analysis and Calculus. At least this was the order in which I swallowed the stuff.

The eight years spent in the lumpily-named Math class were necessary to learn the general stuff you had to master in order to jump into more specialized courses. Learning about defense theory is no different. You need long exposure to basics before you are ready to grapple with specialized topics such as nuclear weapons.

The know-nothings disagree. Many of them believe you can hold intelligent positions on nuclear weapons without much or any exposure to the basics of defense theory. This attitude strikes me as the equivalent of asking a first grader to sit in on a calculus class. Sure, he or she can physically do it. They may even volunteer comments, and whatever they say may make sense to fellow first graders, but it will strike serious students of calculus as inane. The professor will simply roll his eyes and carry on, hoping that the child will eventually grow bored and leave the room.

There is not much use in complaining about the know-nothings and their insistence on their places at the table of public debate. It is the right of every American to remain ignorant on the details of any public issue and yet assert wisdom and insight on the same subject. It is also the right of enough similarly situated people to band together in pursuit of a foolish policy and, whether by voting or lobbying or demonstrating, force the government to adopt that policy. There is only one antidote to this disability and that is education. Education in the issues of defense is crucial because, by definition, without defense the country is defenseless. A defenseless country is already well into the last chapter of its history.

There is a legitimate school of thought that holds that it is acceptable to have no national defense. This is, to repeat, legitimate. A defense policy almost always implies a willingness to resist foes, and therefore usually also means a willingness to fight and shed blood. Some people hold the moral first principle that all resistance is immoral, even that done in self-defense that does not involve violence. Such a view ends the need to go any further in this chapter, except perhaps to understand opposing views.

A second school of thought argues for a defense policy limited to non-violent, or passive resistance. This is not a rejection of the need for a national defense, but the conviction that non-violent forms of resistance can succeed in defending communities and nations. Passive resistance gained fame with Gandhi and later in the civil rights movement in this country, led by Dr. Martin Luther King. Passive resistance translates into demonstrations, sit-ins, and an overarching refusal to compromise with the enemy or to raise your hand in violence against him. Passive resistance has worked in some situations. Those situations always seem to involve an "enemy" who is not really that evil, an opponent whose collective conscience cannot for long tolerate violence being inflicted on those practicing passive resistance. Thus the majority of Americans were horrified by the violence perpetrated against the people in the front ranks of the civil rights movement, and this majority forced the government not only to end the violence but also to enforce the full civil rights of all Americans, including black Americans.

The third broad position is the willingness to resort to force to defend one's self, family, community, or nation. The willingness to use force or to threaten to use force marks American defense policy from the time of the ratification of the Constitution to the present. We have been willing to go to war to defend the country and the country's interests. In fact, we have been willing to go to war even when it was not necessary to defend the country. The nation has resorted to force on many occasions for many reasons.

The big question in this chapter is not whether you are willing to use force to defend the country. This discussion assumes that you are. The big question here is whether the government's first priority ought to be providing for the defense of the country. A number of conclusions follow from the response to that question. The Constitution, by the way, lists providing for the "common defense" as one of its goals, but is silent as to whether it is *the* great goal. There is a crucial difference.

75 Defense

The need for defense can be traced back to the first man who got hit on the head by a rock and lived to think about it. It may have been on Abel's mind when Cain killed him. It is an idea that occurs to every schoolboy who gets pummelled at recess, and to every person who has been burgled or who fears being burgled.

The need for defense quickly becomes apparent to anyone who wants to continue living. The desire for self-preservation must precede any conviction that defense is necessary. If you don't care about living another day, you don't worry about defense. If you do, then you begin to plan.

Student councils do not do much heavy-duty defense planning, though individual students engage in a lot of it. The extortionist in your class no doubt has a list of targets on whom he can put the touch. Chances are the extortionist doesn't demand lunch money from the noseguard on the football team. In fact, the noseguard is a prime suspect for the role of bag-man if he is as unbalanced as noseguards typically are.

The extortionist instead seeks out the tuba player. And the tuba player either forks over, or he begins to plan. He may plan to avoid the extortionist. Or he may plan to be forever within the sight of the Dean of Men. Or he may buy himself a knife. In any event, he gets to thinking about defense in a hurry.

A silly example? Yes, but also not so silly. The confrontation between extortionist and tuba player is a confrontation between two people with opposing goals, at least one of whom is willing to threaten to use force to obtain his goal. The put upon person has the three broad choices available that were outlined earlier. He can pay up—no resistance. He can run away or allow himself to get thrashed again and again—passive resistance. Or he can resort to counter-force or the threat of it. Don't be confused by the variety of tactics open to the tuba player. Hanging around the Dean is, for example, reliance on force, although it is the force of an ally.

Wherever and whenever there is confrontation, there is the need for defense planning. The United States has enemies, though they can be referred to in a more gentle fashion by calling them "opponents." Our first enemy was Britain, but we have had a number of them, including but not limited to Spain, Mexico, Germany, France, Cuba, Japan, Nicaragua, Iran, Iraq, Egypt, Syria, various terrorist groups, the Peoples Republic of China, North Vietnam, Laos, Cambodia, Poland, East Germany, Rumania, and many others. Times change. The list of enemies undergoes a slow but continual evolution. At the close of 1986, our number one opponent was the Soviet Union. The Soviet Union happens to be in an excellent position to threaten the United States and to back up that threat with force. Our situation with the U.S.S.R. does not represent a radical break in U.S. history, because the U.S. has had enemies for over two hundred years. It has always been the policy of the government to defend the country. Only the form and the immediacy of the threat and our response to it has changed, never the conviction that the threat must be resisted.

It is vitally important that you and your contemporaries take seriously the need to carefully assess the threats that face the United States. From the very beginning of recorded history through to the present day comes a single message: Be alert to the possibility of war if you hope to avoid war.

In 432 B.C., representatives of the ancient Greek city-state of Corinth assessed the activities of the government of the city-state of Athens, and concluded that war throughout Greece was inevitable. The Corinthians sought to enlist on their side the only other City with power equal to Athens, Sparta. The Corinthians had been warning Sparta for years that Athens was up to no good, and when a debate on the general subject was scheduled before the Spartan assembly, Corinth sent representatives to argue its view. At least one of these men made a lengthy speech to the assembly. The speech was set down by Thucydides, and began:

> Spartans, what makes you somewhat reluctant to listen to us others, if we have ideas to put forward, is the great trust and confidence which you have in your own constitution and in your way of life. This is a quality that certainly makes you moderate in your judgments; it is also, perhaps responsible for a kind of ignorance which you show when you are dealing with foreign affairs. Many times before now we have told you that we were likely to suffer from Athens, and on each occasion, instead of taking to heart what we were telling you, you chose instead to suspect our motives and to consider that we were speaking only about our own grievances.

The representative from Corinth paused for a bit to state his theory on the strategy the Athenians were pursuing:

> As for Athens, we know their methods and how they gradually encroach upon their neighbors. Now they are proceeding slowly because they think that insensitiveness to the situation enables them to go on their way unnoticed; you will find that they will develop their full strength once they realize that you do see what is happening and are still doing nothing to prevent it.

The Corinthian could not and did not attempt to conceal his disgust at the inability of the Spartans to see the clues —to read the writing on the wall. Wake up, he urged them:

> Think of this too: while you are hanging back, they never hesitate; while you stay at home, they are always abroad; for they think that the farther they go the more they will get, while you think that any movement may endanger what you have already. If they win a victory, they follow it up at once, and if they suffer a defeat, they scarcely fall back at all.

The Corinthian complaint was not that the Spartans did not realize how much power the Athenians had assembled. The complaint was that the Spartans had misjudged the Athenian strategy and had ignored all the clues to that strategy. All of Sparta's considerable power was worthless, they argued, unless it was accompanied by an accurate assessment of Athenian strategy. The Corinthians believed the Athenian strategy was transparent: Athens had prepared for war and war, in fact, had already begun in its preliminary stages.

The Corinthians, it turned out, were correct. War soon engulfed Greece.

The Corinthian complaint has been repeated by various speakers in various countries at various times throughout recorded history. Whether America is now occupying the same position as Sparta is an open question. What is not debatable is the need to gather the facts about the actions of potential enemies and to come to conclusions regarding the nation's response.

Defense planning consists of two broad elements: the determination of strategy and the stockpiling of weapons. Both elements are necessary to an effective defense.

The fellow who got hit on the head with a rock and lived to think about it had immediately to answer the question: "Do I want to live longer?" If his answer was yes, the next question was: "Shall I resort to force in order to accomplish my goal of living longer?" The fellow who runs away is not the subject of this discussion.

Once he resolves to live and to resort to force, the man has to fix on a strategy and get a hold of some rocks. If he's fast, he may want to gather small rocks that he can run around with. If he's fat and slow, he may set up shop in a cave or in hiding and stockpile some heavier stones to use when his enemy wanders into range. Strategy and weapons.

If his enemy has taken a big leap forward in strategy or weapons—that is, if he has taken to climbing trees and dropping rocks, or has come into possession of a Colt .45—man #1 is in big trouble.

Unless your local independent television station is in deep financial distress, you probably have never seen reruns of the 1960's-era series "The Time Tunnel." That's unfortunate, because you will never see the episode where the Time Tunnel dropped an Army sergeant onto the battlefield of Troy. The sergeant quickly sized up the situation, threw in with the Greeks, and hit the forces of Troy with all the force of his machine gun. This created quite a bit of disorder in the ranks of the Trojans and they fell back to regroup. It is also the perfect illustration of the effects of a radical advance in weaponry on an ongoing conflict. So too was Hiroshima. So too might be the Strategic Defense Initiative—"Star Wars." In fact, Time Tunnel only made more vivid the effect of the Trojan Horse—an advance in weaponry—on Troy.

That television episode did not illustrate, however, the equally decisive role of strategy. Had the Army sergeant pressed his advantage and then had his gun jam, he quickly would have been on the receiving end of a lot of spears. Had the Trojan Horse not been accompanied by the Greeks' ruse of elaborate preparations to depart permanently from Troy, it might not have been accepted into the city. Similarly, had President Truman opted to expend only one atom bomb on Japan instead of two, the Japanese might have refused to surrender until the allies had waged a long and bloody struggle for victory on the Japanese mainland.

Strategy requires articulated goals. One agreed upon goal for the U.S. defense planners is to prevent a Soviet victory over the U.S. in a war initiated by the Soviets. If the U.S.S.R. begins a war against the U.S., and the U.S. is not at least capable of fighting that war to a stalemate, then all of our other goals are irrelevant.

So at a minimum, the U.S. needs a strategy that will allow it to survive a Soviet surprise attack and threaten retaliation.

"Wait a minute," comes the protest. "What about a strategy that seeks a peace treaty and thus ends the need for the minimum strategy just outlined?"

This is a good question. Is it a realistic one? Even if such a treaty could be negotiated, does the signing of such an accord mean abandoning the minimum strategy?

Man #1, the recent recipient of a rock to the head, is appoached by Man #2, carrying a white flag. Man #1 is the civil sort, and puts down his rock. Assume that Man #2 is sincere and has not concealed a rock or a Colt .45. (The Trojans assumed that the Greeks were sincere and had sailed away, but forget that for a moment.) Man #2 says that he is sorry, that his rock-throwing stuff is silly, that he no longer has time for throwing rocks at animals and he can't relax. His apology is lengthy and complete. He then suggests that both he and Man #1 get rid of all their offensive rocks.

What should Man #1 do? Suppose that he agrees, and a year later Man #2 hits him with another rock. If he lives, he has serious things to consider--like whether he will ever strike another agreement with Man #2. In the meantime, he's got to run like the devil getting his rocks together.

Recall from an earlier chapter that the Soviet Union signed a non-aggression pact with Hitler's Germany in

1939, and that Hitler launched a massive invasion of Russia two years later. Times change. Paper agreements can be torn up. To repeat, if you are serious about defense, you pray for peace and keep your powder dry.

Things might be different if you could be certain that your one-time enemy has disarmed. Disarmament is theoretically possible, as is its verification. But cheating is not only theoretically possible, it has happened in history time and time again.

There is, frankly, no foreseeable hope that the United States can ever do away with the necessity for defense planning. It would be easy to argue this point by reasoning that even if the Soviet Union and the United States came to terms and set aside their differences, there would still be other nations hostile to the United States. That is, however, intellectual subterfuge. Unless and until the government of Russia is no longer communist, there cannot be meaningful disarmament. Communism has failed miserably to produce either economic growth or social happiness. It depends on military strength to survive. The threat that the Soviet Union will use its military power is out there. It is a constant. Treaties can certainly serve to reduce the need for certain types of weapons and for certain levels of defense preparedness. But they cannot do away with the necessity of at least a minimum defense.

Let's wander even further into reality. The minimum goal of U. S. defense policy is outlined above. What are some other goals that are not in dispute?

We have determined over the years that it is a goal of the United States to defend the countries and the peoples of Western Europe against aggression by the Soviet Union. We will go to war with the Soviet Union and its allies if the Soviet Union attempts to conquer any or all of the democracies of Western Europe. We obviously made no similar commitment regarding Afghanistan. We have, however, entered into similar explicit or implicit agreements regarding Japan, Australia, Israel, Turkey, South Korea, and the Republic of China.

All of these commitments grew out of moral choices made by the leadership of our country, though these commitments were also made with the nation's self-interest in mind as well. This self-interest is also known as "strategic interest." We do not always threaten to go to war when our strategic interests are involved. We also try to avoid peripheral confrontations with the Soviets that can escalate into uncontrollable conflicts. But we have drawn some lines on the globe that we have pointedly told the Soviets not to cross. Having drawn these lines, we are obliged either to renounce them or to plan our defense policy so that our military can realistically be expected to back up our commitments.

A quick illustration is necessary here. Man #1 has struck up a friendship with Man #3. Man #3 is afraid of getting hit by a rock thrown by Man #2. Man #1 has collected enough rocks to feel confident that if he warns Man #2 to leave Man #3 alone, Man #2 will do so. He offers Man #3 his protection, Man #3 accepts, and Man #2 is given warning.

If Man #1 changes his mind about his willingness to protect Man #3, is he under an obligation to tell Man #3? If he says nothing to Man #3, but Man #2 gets wind of Man #1's change of heart, obviously Man #3 is at risk. Or suppose Man #1 lets his supply of rocks dwindle and Man #2 takes note. Has Man #1 violated the trust of Man #3 even though he has not consciously broken with his commitment to defend him?

Man #3 is in a position relative to Man #1 as our allies are relative to this country. We have made commitments to defend our allies. We owe them either an honest declaration that we will no longer serve as their defender or a defense policy that plans adequately for their protection as well as ours.

American defense planning, it appears, must take into account all the commitments this country has made—all its goals—and then add up the men and materials necessary to reach those goals. First, the requirements are assessed for meeting the minimum goal of fighting back from a surprise attack launched by the Soviet Union. Then the requirements necessary to defending our allies are added in. The stuff necessary to reaching our subsidiary goals is then toted up and piled on.

There are some obvious things on the nation's list of subsidiary goals. One of these goals is the preservation of the economic health of the United States. Check out the clothes you are wearing and the goods you are using as well as the car your family is driving. Many of these items have been manufactured in foreign countries. People in those countries are likewise using American-made goods at the same time. America is a great trading country. Hundreds of thousands of American jobs depend upon this trade, and trade depends, among other things, on something called sea lanes—the regular routes of passage to and from American ports that ships travel. The United States is not self-sufficient. Its economic health depends on imports getting here by sea and exports getting out by sea. It depends on open sea lanes. American defense policy must plan on defending and keeping open those sea lanes in times of conflict if the American economy is going to survive and, indeed, if American troops are going to get where they need to go and, once there, receive the supplies they need to survive.

Another subsidiary goal is the protection of Americans travelling or living abroad. Two recent examples come to mind. A group of terrorists hijacked a TWA flight in the summer of 1985, and the Beirut Hostage Crisis began.

Newspapers reported that the use of force to free our hostages was considered at that time. Even though force was not used, the requirements had to be there for the option to have existed even in theory. Force *was* the option chosen when the government, in 1983, noted the threat to American medical students attending school in Grenada, although the safety of the students may not have been the key goal in ordering the invasion. The rescue portion of that operation illustrates that defense policy includes contingency plans for saving imperiled Americans.

The job of defense planners is to figure out how we can accomplish various missions if the need arises. Debates on strategy and the acquisition of weapons follow.

These debates require a serious assessment of which obstacles are likely to be thrown in our path by unfriendly powers. They require as a crucial component an evaluation of the motives and capabilities of our likely enemies. This requires a cold and accurate calculation of the military capabilities of the Soviet Union.

The Soviets possess 9,460 nuclear warheads and bombs that are capable of reaching the United States. They have 1,532 missiles and bombers capable of carrying those warheads to our shores. These figures do not include their nuclear weapons and delivery systems that can reach Europe or Japan, but not the U.S.

The Soviets and their Warsaw Pact allies also have, among other things:

- 6.1 million soldiers
- 50 thousand tanks
- 19,900 pieces of artillery and mortars
- 4,316 anti-tank guided weapon launchers
- 3,150 attack helicopters

They also possess a formidable air force and navy.

Even this small sample of statistics should demonstrate

that our potentially most dangerous opponent has a huge array of weapons. Hardware is, however, only one component of military power. Our defense planners must also assess Soviet military strategy.

The critical question is whether the Soviets are willing to use their forces in support of the Soviet government's objectives. At this writing, the answer is clearly, "Yes, at least in some situations." Soviet armed forces are waging a brutal war in Afghanistan. There are over 100,000 Soviet troops fighting this war, and the war has gone on for more than eight years.

If the Soviet government is willing to use force in one part of the world, that does not necessarily mean it is willing to use force somewhere else. The fact that the Soviets invaded a country on their border does not mean that they will invade a another country somewhere else. But the Soviets have also been very willing supporters of the military adventures of their allies in Vietnam, Ethiopia, and Angola. Soviet troops invaded Hungary in 1956 and Czechoslovakia in 1968. The Soviets are obviously willing to use military force in many situations.

So too is America. Since 1945, our troops have fought in full-scale wars in Korea and Vietnam. U. S. Marines have landed in the Dominican Republic, Grenada, and twice in Lebanon. We supply weapons to other governments and to non-government guerilla forces.

Which brings us to the central problem in the contemporary debate about defense policy. Both the Soviet Union and the United States use military force. Is there a moral difference between the U. S. use of force and the Soviet use of force?

A poll was recently taken in a high school very near Washington, D.C. Fifty-three high school seniors were asked whether there was a "moral difference" between the U.S. and the U.S.S.R. Fifty-one of them—that is 96 percent —said no.

Is there a difference between sponsoring an invasion of South Korea and helping the South Koreans defend themselves? Yes, of course there is. Is there a difference between fighting to keep the people of South Vietnam and Cambodia free from the rule of the North Vietnamese, and providing the weapons to the North Vietnamese by means of which they invaded and conquered South Vietnam and Cambodia? Yes, of course there is. Is there a difference between invading Grenada and leaving, and invading Afghanistan and staying? Yes, of course there is. These are the differences between the U.S. use of force and the Soviet use of force. It is a profound moral difference.

The most difficult inquiry along these lines is whether or not there is a moral difference between supplying arms to the Contra rebels in Nicaragua and supplying weapons to the Sandinista government of Nicaragua. The U.S. does the former and the Soviets do the latter. Yes, there is a difference. The Sandinistas are a communist regime that has already brought repression to the people of Nicaragua and, if unchecked, will attack its neighboring states. The Contras want to overthrow the Sandinista tyranny and replace it with democracy.

The distinctions between the U.S. and the U.S.S.R. must inform serious defense planning: The Soviet Union is an agressive, expansionist power while the U.S. is a defensive power.

Because the U.S. is willing to accept the world as it currently is and is not willing to launch any wars to reclaim for freedom's sake any of the territory now firmly under Soviet rule (East Germany, Poland, Hungary, etc.), we do not need the weapons and strategy necessary to roll-back the Soviet Union. We do, however, need the weapons and strategy necessary to deter an attack by the Soviets on the U. S. or Europe or places such as Israel.

Should this goal of providing an adequate defense be the first priority of the government? The government has only so much money to spend.

If an adequate defense is the first priority of the government, then the first question the government will answer is: "What weapons and how many men do we *need?*" The size of the defense budget may astound you. You may wish that all that money could be spent on providing the best possible housing for the elderly or three hot meals a day for impoverished children. But if defense is the nation's primary need, then the first thing the government does is obtain the materials and men necessary to defend the country.

The issue of exactly what we need is beyond the scope of this book. The government has decided, for example, that by 1990 we will need 15 aircraft carriers and a total of 600 warships. Why? Because it is 4,900 nautical miles from San Francisco to Korea and 5,100 nautical miles from New York to the Suez Canal. These are only two of the many sea lanes the Navy must be prepared to defend in order to defend the country, and the experts have determined that we will need 600 ships (617 to be exact) to do that job in 1990.

I hope that you will by now have come to understand the general process by which the government arrives at the 600 ship figure. Defense analysts count up the number of Soviet ships and the ships of Soviet allies. They pay attention to the number of submarines. Then they assess the capabilities of these ships and the threat they would pose to U. S. supply and trading vessels if war were to break out. Then they calculate how many ships we need in addition to the fleets of our dependable allies. That number turns out to be 617 in 1990.

People can and do disagree about the particulars. Perhaps 550 ships would do the trick. Maybe 700 are really necessary. I sure don't know and I have to rely on people trained in the art of defense analysis. But when those people come up with their requirements, I think we ought to build those ships or repair the ones we have so that 617 are afloat in 1990. And we should do that before we pay to keep Amtrak running between New York and Washington or build a canal in Louisiana.

Too much time is spent arguing over the particulars of defense procurement by people who have not given enough time to the general questions associated with a decision to defend the country and the priority that ought to be accorded to that goal. The particulars make sense only when you have a good grasp of these basic questions.

People love to debate nuclear arms and the merits of particular nuclear weapon systems. The United States spent over 313 billion dollars on defense in 1986. That was slightly over 7% of all the money and goods produced in this country in the same year. Of that 313 billion, about 30 billion or roughly 10% was spent on nuclear forces. To discuss the defense budget is to discuss only a small portion of the nation's wealth, and to discuss nuclear weapons is to discuss only a small portion of the defense budget. End spending on nuclear weapons tomorrow, and the national budget deficit stays in the 140 to 170 billion dollar range. That cut-off would not mean more money for the poor.

Nuclear weapons do not, of course, command so much public attention because of what they cost, but rather because of what they can do. What they can do is kill millions of people in an instant. What they also do is defend millions of people every day.

As I mentioned earlier, the Soviets have 9,460 nuclear warheads and bombs that can reach the United States. The United States must consider that threat and decide what is necessary to deter the U.S.S.R. from launching those weapons.

Nuclear weapons pose a tremendous threat not only because of their destructiveness, but also because of the speed with which they can reach their targets. An all-out Soviet first-strike against the U. S. would be over in about thirty minutes. Missiles targeted on Washington could, in some cases, hit there ten minutes after the order to fire is given.

So how many nuclear weapons does the United States need to deter such a first-strike? Use your intuition on this one. Assume that every American nuclear weapon can destroy Moscow if it gets to Moscow.

Would one such weapon be sufficient deterrence? The Soviets can hit the U. S. with, at a minimum, a few hundred warheads at once. Would one U. S. bomb deter such an attack? Would ten, or 100, or 1,000?

Well, it depends. It depends on whether or not, following a first-strike, the U.S. would have any way of delivering a nuclear counterattack. And it depends on the Soviet defenses, such as planes and missiles that would try to shoot down incoming bombs. (Moscow is surrounded by an operational Anti-Ballistic Missile System. The U.S. does not have any such system around Washington.) Most important, it depends on what the Soviet government would consider to be an acceptable level of casualties and damage. This is a hard thing for U.S. defense planners to judge. They do know, however, that the U.S.S.R. absorbed 20 million casualties during World War II and never surrendered or accepted anything less than the total and unconditional surrender of Germany.

So U.S. nuclear planning depends on figuring out how many weapons we need so that we can inflict an unacceptable level of damage on the Soviet Union after a Soviet suprise attack. That many nuclear weapons equals deterrence.

By now you should be convinced that nuclear math is difficult stuff. One final thought should be added to our calculations. Since defense planners are dealing with an extremely dangerous threat, should they arrive at a minimum number and variety of nuclear weapons, build those, and then be content; or should they err on the side of safety and build more than that which they calculate to be the minimum?

Once you have answered these basic questions, you can understand the purpose of the nuclear arsenal of the United States. You can then understand why that arsenal takes the form that it does. Our strategic weapons—those that can reach the Soviet Union and carry a big bang—are divided into three types: land-based missiles, submarine-based missiles, and nuclear-armed bomber aircraft. This is the strategic "triad." The nation depends on the triad in order to make it more difficult for the Soviet Union to launch a successful surprise attack. Even if the U.S.S.R. developed the perfect killer weapon for use against our land-based missiles, it would still have to possess the capability to find and destroy our nuclear-armed bombers and our nuclear submarines.

The need to make our nuclear forces as invulnerable as possible to surprise attack also explains why the defense planners are continually trying to develop more accurate weapons—so that fewer weapons will need to survive in order to inflict an unacceptable level of damage on the Soviet Union—and weapons that are easier to hide.

Opponents of modernization like to argue that we are egging the Soviets on, and that the Soviets fear our advances. They say that we are fueling the arms race.

This country flat-out stopped building Minuteman missiles in the early 1960's. The production lines were shut down. We had a huge lead over the Soviets in numbers and accuracy. The Soviets caught up, and have subsequently built many new types of more accurate and more powerful missiles. We are now trying to catch them. They have never demonstrated the willingness to accept equivalent forces on both sides.

What about the old refrain that we have enough weapons to blow the world up ten times? Perhaps we do if we just started setting off weapons all over the world. But no one imagines doing that. The real question is: Do we have the kind of nuclear arsenal that convinces the U.S.S.R. not to attack us or our allies? If they launch an attack on us and destroy all our arsenal, we could not blow up Rostov much less the whole world.

Nuclear weapons are not going to go away. We have 11,500 nuclear warheads and bombs in our strategic forces, and 1,993 missiles and planes to take those weapons to their targets. Is that enough to deter a Soviet surprise attack? It was in 1986. It may be in 1987. But currently our only defense against a Soviet surprise attack is the threat that we would hit back with devastating force. That threat must be convincing to be effective, and the people who need to be convinced are not the Americans—they are those in control of the Soviet government.

If you believe that defense is the first priority of the government and if you accept the idea that our nuclear arsenal is necessary to make that defense work, you will not be misled by slogans and simplistic appeals to emotion such as those of the Nuclear Freeze Movement. You may still become a harsh critic of current defense planning, but only after you have asked yourself whether or not we have a balanced and effective nuclear arsenal and whether or not that arsenal in combination with our conventional forces and the forces of our allies can deter an attack from the Soviet Union. You will get around to examining the particulars.

By now you are probably as tired of this subject as I am of trying to organize the basics on paper. This chapter is the longest in this book for a good reason. It is neither the most difficult chapter to understand nor the most important. But it is in this area that the illnesses of contemporary thinking have taken their greatest toll. This is the subject that has attracted far and away the greatest amount of bad thinking. It is on this subject that many people will urge you to take short-cuts when there are no short-cuts to getting a grip on the issue.

POVERTY

First, some statistics:

- As of September 1985, 33.7 million Americans were living below the government-defined poverty line: an income of $10,609 a year for a family of four. That means 14 percent of all Americans are living on less than that amount of money, some on a lot less.
- At the start of 1984, 2 million Americans were classified as "long-term" unemployed, double the total at the start of 1982.
- During 1983, there was a 71 percent increase in the number of Americans resorting to emergency feeding programs in the twenty cities surveyed.

Next, two comments:

- "In 1966, the poorest fifth of American families had four percent of total American family income before tax, and the richest fifth had around 47 percent. Two decades later, the figures are almost exactly the same. In other words, all the programs of the Great

Society, all the welfare, food stamps, increased Social Security, and the like have simply kept things in place. If the programs had not existed, the gap between rich and poor would have increased enormously."

—The Editors of The
New Republic,
February 18, 1985.

- "We have learned enough over the past two decades to acknowledge that hundreds of thousands of Americans are so short of money that they are constantly hard-pressed to obtain an adequate diet—as well as other basic necessities."

—Author Nick Kotz,
"The Politics of
Hunger"
The New Republic,
April 30, 1984.

Finally, one anecdote:

- "No one seems to know anything about Warren E. Wooten. At a homeless shelter about a half-mile from the Capitol, he sits in a chair talking to himself. I ask Mr. Wooten why he's here. In agonizing, slurred sentences, he says he was a doctor and a general in the Army when he left in 1960 due to a 'neurosis.' (He later tells me that after he left the Army he was 'arrested and sent to St. E's'—St. Elizabeth's Psychiatric Hospital.) Now, he says, he is a U. S. Supreme Court Justice. Mr. Wooten gets one meal a day at the shelter—and that 'not fit for swine.' He was receiving $1,103 a month in disability when he was living with his brother. But since his brother left town, and he was evicted from his apartment, he hasn't gotten a cent. Volunteers for the Community for Creative Non-Violence, a homeless rights group which runs the shelter, say that many of the seven hundred people they see there each day have stories like Mr. Wooten's. Most have had some history of mental problems. Food is available at churches and soup kitchens throughout Washington, but help in transporting all those people who need the food isn't.

—Jonathan Rosenblum
" 'Anecdotal Stuff' "
The New Republic
April 30, 1984.

The statistics are there to alert you to the scope of American poverty. The comments are intended to deflate the seductive notion that things are getting significantly better as the years roll by. The anecdote ought to give you a quick glimpse of poverty in terms of individuals.

Most Americans, myself included, have effectively immunized themselves from the reality of poverty. Everyone will willingly acknowledge that an awesome number of Americans are grindingly poor; feel hunger frequently, if not every day; and have never experienced even the mundane pleasures of owning a bike or going to a movie. Some of those who will acknowledge this state of affairs also are quick to point out that there are loafers and cheaters among the folks officially classified as poor who really are not that bad off.

Poverty in America is another one of those issues surrounded by bad thinking. You have to fight your way past dozens of side issues and irrelevant comments even to approach the central difficulties posed by poverty. One of the biggest obstacles is the one just mentioned: the idea of freeloaders populating the welfare rolls. Who cares, you can answer. If, as almost everyone is willing to do, you acknowledge that there are large numbers of truly impoverished people in the United States, then all the cheaters and loafers are irrelevant to the big issue of poverty. If you are prepared to think through the problem of poverty, be prepared to recognize a lot of irrelevant debates, and to push them aside.

Another difficulty in thinking clearly about the truly poor is that this group almost always is the captive of someone's agenda. Whenever this group makes news, spokesmen for the Left and Right are quick to offer editorial comment and interpretive analysis.

One example. In August of 1985, the Commerce Department announced the most recent data on poverty in the United States. The number of Americans classified as officially poor had dropped between counts from 35.5 million to 33.7 million. Within two weeks the Washington Post ran side-by-side commentaries on the significance of this drop. An Administration official reviewed the numbers, mused about some additional steps that could help keep the total dropping, and concluded that "[w]hen it comes to welfare-state approaches to poverty, the liberals are intellectually bankrupt." Across the page, the observer from the Left blasted the "hypocritical puffery" of the Reagan Team, and concluded that the "challenge is to find political leaders who will make a real effort to do more about the one in seven Americans who were poor in 1984." Both commentators were using poverty statistics to support political premises.

The manipulation of the plight of the truly impoverished is not new, and it will not end anytime soon. The objective facts of poverty will always serve as ammunition in political wars the ultimate ends of which are the capture of political power.

Poverty is easy to manipulate because it is far removed from most Americans. Further, the truly poor are not in much of a position to speak for themselves. They rarely read much less comment upon the abundant analysis of their situation. If there is one truly poor person who reads this book, for example, I will be surprised.

97 Poverty

A remote problem lends itself to abstract solutions. Real reform, and the great political upheaval that gives rise to such reform, typically accompanies widespread agreement and feeling about particular grievances. The Civil Rights Movement of the early 1960s was so startlingly successful because it sought to change a set of circumstances about which most Americans were aware and ashamed. But poverty is not that kind of issue. Poverty has always been around, and will likely always be around. The very quality of inevitability that surrounds poverty renders the issue a dull ache that is endured rather than a sharp pain that requires radical treatment.

This chapter on poverty is the last one I wrote, though not the last to appear in the order of this book. I wrote it last because I know next to nothing about poverty. I'm rich, though not in the eyes of wealthy folk. (I pull down $70,000+ a year. The stats say that less than 3 percent of people my age earn such a figure. So I figure I'm rich.) I know nothing about poverty, and the odds are that neither do you. You and I may both have felt the sense from time to time of being pressed for more money, but that sense is light-years removed from feeling poverty.

When I lived in California, I took a day trip to Tijuana, Mexico, and unintentionally strayed from the tourist's path. The slums of that city are revolting. I saw thousands of families living in refrigerator boxes and similar shelters. Even at this distance, the power of poverty to radicalize people is apparent. The injustice is obvious. Once poverty is seen even at arm's length, people must either alter their ideas about poverty or must block the reality of it from their thinking.

The latter choice is one that is made every time the TV runs an ad appealing for donations for famine relief. The pictures of children with distended bellies and shriveled limbs are not appealing, and often the reaction of the viewer is a quick channel change.

I do not think that this is an abhorrent reaction. Not everyone has the capacity to work among or even look upon the poor for very long. This inability is not deplorable. It is understandable.

At the same time, the great silence concerning the truly poor and the conditions of their suffering is not something to be swallowed as inevitable. I take for granted your answer of "yes" to the question, "Ought the poor to be helped?" The big question in this chapter is how rationally to do the helping.

It's two questions, really. How does an individual respond, and how does a government respond?

Every American has around him or her a hundred opportunities to do something to alleviate the immediate impact of poverty. I do not mean sitting around and thinking up a better poverty policy. Good ideas are needed, of course, but their development and debate is no substitute for paying attention to the often grimy business of helping out one single poor person. You do not have to work as a volunteer in a soup line once a week. As I said earlier, some folks are not cut out for that. But you can always give money to the folks running the soup line and taking care of Warren Wooten. That's simple and effective. If you do not involve yourself at even this pristine level, your credibility in the debate on poverty is zero. Even as a high school student, you, today, possess this capacity to help alleviate the immediate impact of poverty. All you need is an envelope, a stamp, and a dollar bill. American generosity towards the poor is deep, but it is also narrow. Too few people are involved. Many confuse philanthropy with charity, a necessary but obscure distinction. Many organizations that do good things do nothing about poverty. Museums, for example, do good things but do nothing about poverty. Many individuals similarly do good deeds by the dozen, but often these deeds have no connection with poverty. All sorts of groups and people ask for financial help, but few of them need food or shelter to survive. Clear thinking about poverty demands that efforts to aid the poor not be lumped in with all other types of private philanthropy.

What ought the government to do? The Constitution provides that one of its primary goals is to "Promote the General Welfare." Certainly within that goal is included the obligation to promote the general welfare of the poor. The key inquiries are, "How high a priority is attached to that goal," and, then, "How best to reach that priority?"

I do not believe that promoting the general welfare of the poor can be the government's highest priority. The previous chapter argues that the defense of the country and its Constitution is the first priority. But defense spending currently consumes barely a quarter of Federal tax revenues. Three-quarters of every tax dollar is allocated to different programs.

The competition for non-defense tax dollars is intense. There are dams to build, Amtrak trains to run, research and development grants to make. How ought the claims of the poor to be evaluated alongside these competing demands?

My personal conviction is that the requirements of the poor ought to be second in line, right behind the requirements of national defense. This means that I think we ought to take care of the truly poor before we write out checks for any other non-defense cause. I believe in the truth of the old saw that a country is best judged by how it treats its worst-off.

A coherent poverty policy would aim to raise those at the bottom of the pile to the next level up on the scale of destitution, and when that has been done, to raise that entire second level to the third, and so on. The money will run out somewhere, but the worst suffering will also have ended.

This may sound rational to you, but it never works this

way. Once a program is in place and functioning, it is nearly impossible to end it. Money is sucked up into established programs year after year despite the fact that the most destitute people seem to remain that way year after year. There are lots of reasons for this state of affairs. One of the more sophisticated explanations is that any program, once it gets started, quickly develops a constituency that includes sponsors, administrators, and beneficiaries who defend the program for obvious reasons of self-interest: ego, jobs, and benefits. A simpler explanation is the fear of the unknown. As the old saying goes, man prefers a hundred devils he knows, to the one devil he does not know. A third explanation is the fear that the promise of new approaches is only a smoke screen, the cover of which provides an excuse to kill a program.

No matter which explanation is valid, there remains a startling inability and unwillingness to experiment in the area of governmentally sponsored poverty programs. The fear of being charged with "insensitivity" has made Congressmen unable to make distinctions among the poor. This inability seems crazy. But it's there.

The specifics of public policy demand that you always remember the necessity of making distinctions. Imagine a line of ten street people. Armed with a dollar to donate, you survey the ten people in the line. You can make change and give every one of them a dime. Or you can judge who are the two worst off, and give each of them fifty cents. If one of them strikes you as particularly destitute—say one has no shoes and it's winter—perhaps you will give the entire dollar to him or her.

If you come across the same line the next day, again with a dollar to give, and another person has joined the queue you must rethink your choice. There is always this process of re-evaluation of how best to distribute money, a process the government denies itself except on rare occasions.

Obviously we are talking about tough choices. Anyone who walks up Connecticut Avenue in Washington, D.C., on any given day makes such a choice if he gives a few coins to one of the city's many street people. If individuals can make choices based on distinctions, government can do so as well. In the area of poverty policy, it must.

The point of this entire chapter is to alert you to poverty in America, to urge you to consider its place among the government's priorities, and to suggest that it is wrong to be imprecise in targeting the beneficiaries of poverty programs. Every badly spent dollar robs someone on whom it would be better spent.

ETHICS

If you are a student in a Catholic high school, you have undoubtedly seen her, and may have been taught by her. In your mind she appears to be about 90-years-old. In fact, she is closer to 65. She is short, sharp-tongued, and accustomed to dozing. She will not retire, so the school authorities task her with teaching the religion class for ninth graders. For me, she was Sister Georgeanne. For you, someone else. Looking back, I realize she was a great teacher. Not that her lectures were always original. In fact, she had only three, and those three were repeated during the entire year. Lecture A was on the dangers of sex. Lecture B was on the idiocy of drinking. Lecture C addressed the "all other evils" category. Drugs were not a part of her experience, so they did not enter her lecture sequence. Her counterpart of today no doubt has added Lecture D to deal with drugs. Sister Georgeanne was a great teacher. She knew what her students wanted to be informed about.

I have taught a class of high school students in the Confraternity of Christian Doctrine Program ("CCD") for nearly four years: two years in California, a year-and-a-half in Washington. This once-a-week class is supposed to be a religion class for Catholic high school students attending public schools. It was never that. Oh, I tried to talk about religion, to teach about the Church and its history and doctrine, to focus on Christ, and to discuss many of the issues touched on in this book. But the students did not want that. They wanted to hear about, and occasionally discuss, sex, drugs, and rock and roll.

What's a teacher to do? These are issues of immediate consequence to high school students. No doubt some of my students used drugs. Perhaps some had also had sex, and had undergone abortions as well. All of them listened to music I found incomprehensible, even though I was not much their senior.

I know that lurking below my students' not-so-subtle efforts to direct class discussion to these subjects was a great mass of personal experience. How can anyone not know that? Teenage pregnancy in this country is, for example, epidemic. One million teenage women become pregnant in this country, *every year.* Close to half of them have abortions. While the explosion in pregnancy is relatively recent, high school students have been drinking since there was beer and wine, and drugs are plentiful, cheap, and available. It is no accident that automobile mishaps are the leading cause of death among teenagers, and that a huge number of these crashes are alcohol-related.

The routine statements teachers make about ethics can be repeated from memory by most seniors in high school. These statements are usually not very influential. Influential opinion on these subjects is developed from within. You will call the shots for yourself. Now you are fettered only in that you must be mildly ingenious in order to buy booze, obtain drugs, or arrange to be alone with your current love. Even those inconveniences will vanish next year. They will be replaced by a lack of money or an inability to arrange a date. There will be, in some instances, a continual fear of law enforcement authorities, but on most campuses that is not much of a threat. We are talking about a nearly complete freedom to do what you want with whomever you want without fear of reprisal unless you try it with someone who does not consent. If everyone involved is game, the game is on.

Sounds good, doesn't it? My roommates and I celebrated liberation with a great quantity of cheap wine and Colt .45 beer our first night in the dorm. Things do not change that much.

So what should a teacher tell his or her students on these subjects, and an author his or her readers? Only that a very wise man—the smartest of his age—gave this advice: Moderation in everything. A more modern man of my acquaintance, also very wise, appended this corollary: "Including moderation." This is good and sound advice for someone with no other training in ethics.

I have never met anyone under the age of 18 who was truly ruthless, anyone who truly enjoyed causing serious injury to others. I have met a lot of teenagers who were blind to the pain they quite clearly inflicted, or who were indifferent to causing other people inconvenience as opposed to pain. But a truly evil 18-year-old is not in my experience, although he or she is certainly out there.

Because of this experience I believe that most teenagers are open to being taught a rudimentary system of ethics, and then following it. This system, used in all the classes I've taught, comes down to asking in one way or another, the same question: "What are the likely short term *and* long term consequences of doing this?" The answer almost always comes out as some variation of: "Pleasure and, possibly, some pain." The follow-up questions should be obvious: "How much pleasure, how likely the pain, and who will suffer if the pain does result?" The third inquiry is simple: "Is it worth it?"

"It" may be worth it. "It" frequently is. There are tradeoffs in every decision to pursue pleasure. My point in class, and here, is not to urge particular decisions, but to urge younger people to assess these questions from time to time.

I do not suggest that you paralyze yourself before taking every sip of beer. That's stupid. But the pursuit of pleasure has many forms. Sex, booze, and drugs are three. So are smoking, running, and weight-lifting.

Every category of pleasure you pursue can be examined occasionally. You can figure out how much pleasure you are getting, how much pain you are causing, and who is on the receiving end of the pain. And you can ask if it is worth it. You can adjust your behavior to reflect your conclusion. My own approach nowadays is to shy away from those pursuits that are moderately likely to cause even small amounts of pain in other people. Perhaps you will be comfortable with such an approach.

Sister Georgeanne knew what she was about. She was cranky, sometimes confused, and, to us, hopelessly archaic. I am convinced, though, that she knew quite well that she was the teaching equivalent of a broken record. Turn a subject over and over again before a group of students, and, by force of repetition, many in that group will come to see most or all of the facets of a dilemma. Sister Georgeanne was not candid regarding all of the pleasures involved in the issues she discussed, but she really did not need to be. She was more concerned that we glimpse the consequences of choice.

These consequences do not seem to receive much, if any, consideration from students your age. At a distance, adolescent behavior appears almost programmed, almost beyond the exercise of free will. The recent surge in attention paid by parents and teachers to the ability of students to make intelligent choices is an effort to combat lemming-like behavior in your age group. The now famous "Just Say No" campaign is inextricably bound up with the idea that students do possess an enormous "choosing" power. It is an effort to alert you to that power and to argue that it ought to be exercised wisely.

I am sure you agree with the observation that parents and teachers share a fairly uniform set of views on which options are terrible ones for students to choose. They don't want you to use drugs, to drink, much less drive after drinking, or to climb into bed with your steady. In past decades, this consensus operated to keep students far away from the agreed-upon menu of stupid choices. I do not believe that the consensus has changed all that much. But your ability to avoid the roadblocks that authority puts in your way has skyrocketed. Drugs are available in every community. Fake IDs are as well. Some schools are even in the business of dispensing contraceptives. Parental tactics have been forced to change.

The arguments against destructive behavior and in favor of ethical behavior can no longer be delivered in a peremptory tone from on high. With the erosion of effective techniques to discover and punish prohibited behavior, the need to convince you has risen dramatically. Time and effort must be spent persuading students that older people have excellent, tested reasons for condemning certain choices. Even when a sincere effort is made to persuade it often fails. It fails, I think, because a young adult typically possesses great confidence in his or her own ability to try everything without becoming a hostage to anything. The readers of this book are peculiarly possessed of this sort of confidence. Many of you could no doubt sit through a thousand lectures on the dangers, indeed the idiocy, of mixing beers with cars and still say to yourself at the end: "True enough for most people, but not for me." If the lectures you hear stress only the harm you may do to yourself, they will for this reason often fail to change your thinking one bit.

This chapter tries to drill home a completely different perspective. It allows you to assume that you are in fact invulnerable to the dangers of what have been demonstrated rather convincingly to be destructive choices. Good. You are different and your abilities are superior. I'll buy that. But no matter how talented you may be, it is an inescapable fact that your actions necessarily affect--in a dramatic way--both that group of friends and family to whom you are quite close and an astonishing number of perfect strangers. Every decision you make ricochets off a huge number of people, sometimes with devastating effects.

"Proximate cause" is a familiar concept in the law of injuries. When someone is hurt, the law begins a search to find the responsible party. Thus if Spot takes a chunk out of Smith's leg, Smith can usually recover the cost of medical treatment and a new suit from Spot's owner. If Jones' car runs into Smith's and Smith's as a result is forced to hit Doe's, Jones pays Smith *and* Doe.

As a fairly bright individual, you ought to be able to calculate the repercussions of your actions. Most of the choices that older people suggest are stupid carry with them adverse effects for others. Drinking and driving, for example, kills people. The apparently innocuous purchase of a small amount of drugs sends money into the hands of the most ruthless set of thugs around. These aren't debatable propositions. They are facts. Account for them.

The simple questions I've sketched, even if consulted only once or twice during your college years, will place in stark relief the bodies you have walked over or merely collided with. Not a cheery sight, by any measure, but a view you need.

This scheme is a rudimentary one, a beginner's course, because a higher system of ethics would concentrate your mind on your own character. Eventually you may graduate to such a system. In the interim, however, you'll be saving an enormous number of people a lot of grief if you occasionally trace the patterns of your own ricochets.

RELIGION

The Ayatollah Khomeni is a religious leader. Billy Graham is a religious leader. The Reverend Moon, the Reverend Jerry Falwell, and the Reverend Louis Farrakan are all religious leaders. So is Pope John Paul II. So is Rabbi Meir Kahane. So were Mohandas Gandhi and Martin Luther King. So is Mother Theresa. All of these people are religious leaders. Obviously the description "religious leader" is a broad one. Throw out that net and you are likely to haul in an odd collection of folks.

What are we to make of the various religions that give rise to these religious leaders? Few questions are so difficult to answer or so likely to start wars. Only in recent decades has ideology replaced religion as the driving force behind the most violent conflicts. Today, religion is making a comeback as a source of shooting, as the Middle East, the Iran-Iraq war, and the strife in Northern Ireland attest. Obviously, religion, and the passion it ignites in its adherents, requires a lot of hard thinking.

Since religion unleashed can be so dangerous a thing, we have to ask, "How much freedom ought to be allowed religion?" This strikes many Americans as a ridiculous question, because our country is so accustomed to thinking of itself as a nation with a deep tradition of religious liberty. That is a partially deserved reputation. Americans are absolutely free to believe in whatever religion they want, period. But no American is absolutely free to *practice* any religion he or she chooses. In fact, the practice of religion is quite tightly circumscribed.

Most law students run into these limits early in their legal education because judges have written a good chunk of constitutional law on the subject of freedom of religion. One religious group fervently believes in the necessity of handling poisonous snakes. Not surprisingly, some of the snake handlers are bitten and die. Should the state step in to end this practice? Another religious group firmly believes in the healing power of God, and refuses much of the help that medical technology offers, such as blood transfusions. One believer has a child who will die without a transfusion, and the parent denies the doctor's urgent request to be allowed to give the child the blood needed. The doctor then rushes off to a court with the plea that the judge order the parent to stand aside.

Both situations are real cases. If religious freedom was absolute, the first case would be easy, and the second only slightly more complicated because the religious belief of the child was being asserted indirectly through the parent. But in both cases the courts ruled against religious belief and circumscribed religious freedom. Snake handling was banned and the transfusion was administered.

These examples are out-of-the-ordinary, but they illustrate the recurring problem of church colliding with state. Every religion has its run-ins with government authority. The respective arguments and rights are never easily reconciled.

We could make all of these problems easy by simply

allowing lawmakers to pass whatever laws they wanted and to establish a Federal Department of Religion. But the Framers of the Constitution ruled that out in the First Amendment to the Constitution, which reads, in part:

> Congress shall make no law respecting an establishment of religion, or prohibiting the free exercise thereof . . . .

Great, you say, what does that mean? And doesn't that Amendment only apply to Congress anyway—that's what it says?

A first encounter with Constitutional Law can be confusing. This portion of the First Amendment contains two clauses, known as the "Establishment Clause" and the "Free Exercise Clause." People who have stayed up late in the library for years on end still puzzle over what these two clauses mean. Long ago the courts decided that they applied in full force to state governments as well as Congress, but that decision did not address the central difficulty of the Amendment's meaning. These two clauses are in many respects irreconcilable. Lawyers dress up this irreconcilability by calling it "tension." By that they mean that the two clauses are like two equally-matched teams engaged in a tug-of-war. Sometimes Team A wins, sometime Team B.

Some controversies yield obvious results. The government cannot tomorrow pass a law that outlaws the Reverend Moon's Unification Church because the government is barred from prohibiting the free exercise of any religion. A second easy case: The Congress cannot tomorrow declare the Roman Catholic Church to be the official church of the United States and then start to send tax monies to its parishes in the U. S. That would be a prohibited establishment of religion.

Most questions in this area fall between obvious infringements on free exercise and obvious establishments of religion. A student group requests the use of a public school classroom in which to hold lunch-time prayer meetings. The school authorities say no. The school authorities are the government, a small division of it to be sure, but part of the government nonetheless. What did the government do here? Did it avoid giving support to a particular religion, thus avoiding an unconstitutional establishment of religion? Or did it interfere in the free exercise of the students' religion? Not by any means an easy question to answer.

Another example that you should be used to by now involves the situation where a city government sets up a manger scene in the public park at Christmas. Is the city giving impermissible official endorsement to Christianity, an impermissible establishment of religion? Astoundingly, the Supreme Court has given an answer that seems to be: No, the city *is* allowed to do that; but there must be a Santa Claus close by the manger, and we'd prefer that private funds be used to pay for the entire package. I use the description "astounding" not because the answer is clearly wrong, but because all of these qualifications are supposed to be found in the sparse language of the First Amendment. There are two hundred years worth of court decisions on what the First Amendment means, the collective message of which is that nothing is as simple as it seems.

I will not lay out even a summary of the path the Supreme Court has followed in interpreting the religion clauses of the First Amendment. It defies easy summary and is tough enough going for law students. The values that compete against one another in these controversies, however, can and should be understood.

On one side is history's most powerful and enduring

force: the urge of like believers to worship together without the interference of others. Religion has been traditionally viewed as a tremendously good thing. Most members of our government and the vast majority of Americans practice one form of religion or another. And most Americans want the government to treat all religions alike while at the same time believing that the government should not interfere with the apparently benign practices of religion in public—such as prayer in public school and Christmas creche scenes in public parks. Many folks participate in or at least applaud the great and good things that churches do, and they want the government to facilitate those good things. Church-backed schools, for example, cost a great deal of money to operate, and the churches pick up the tab. Some ask, "Would it really hurt if school children on their way to church-run classrooms hitch rides on the city school buses?" The parents of those students pay the taxes to buy these buses, the argument runs, so why not allow their children to share in the ride? There is no demand here for an independent expenditure of money to build buildings or pay the salaries of teachers in religious schools. That would require additional tax monies. But the buses are out there and have been paid for. Why not allow religious schools to use them?

On the other side is a real fear that if the government gets involved in even apparently harmless projects such as sponsoring a nativity scene, it immediately sends an unmistakable signal as to which religion is the preferred faith. Along with that signal is a second one that quite clearly indicates to non-Christians that their religions, or their lack of any religion, is second-class stuff.

This set of problems scrapes at our sensitivities. Other religions injure our tolerance.

Some religions have come to be known as cults. The Indian guru, the Bhagwhan, who set up a commune in

Oregon, was branded a cult leader. Jim Jones led a different cult. There are scores of others. Most are accused of brainwashing young adherents, some of stockpiling weapons, or plotting to kidnap officials and blow up public buildings.

Still other religious leaders have begun routinely to thrust themselves and their followers into political debates already surrounded by too much heat and too little light: nuclear arms, abortion, Central America. Some have mobilized voting blocks and attempted to influence elections. All of these efforts and the controversies they spawn have given rise to much talk and some muttering about the proper place of religion in America.

Answers to questions in this area are few and far between. Even when one problem appears resolved, the next quickly pops up and often turns your thinking on its head. This lack of steady consensus results because most religious issues are addressed *ad hoc* without resort to the "why" behind the Framers' dual guarantees of no "official" religion and no interference with any shade of religious belief.

How much religious liberty can the country stand? We have agreed to accept, and indeed celebrate, a near-absolute freedom of the press. But is the American consensus still strong regarding freedom of religion, especially "fringe" religions? Don't you think that most Americans have, at one time or another, wondered why we just don't bulldoze some of the far-out religious settlements and whisk the cult leader away? Aren't you certain that most Americans were profoundly pleased when the Bhagwhan was given the boot?

Of course they thought and felt these things, but there is also a voice, at least in the minds of considerate people, that says: "What about my church, or my child's church? Who gets to direct the bulldozers?" If your answer is that the government should have its hand on the throttle, that should not dampen your apprehension, because history shows that repression's thirst, once whetted, seems inevitably to become unquenchable. Just look at the world. Violence that stems from an intolerant belief that one religion is *the* religion claims many lives. Do we want to mess around with the basic tolerance that has made America immune to the worst forms of this violence?

The parents of a cult follower do not care much for this business of tolerance. They may admit that the dangers of intolerance are real, but add that so is their dilemma. They look at their kid and they judge that he or she is trapped. They remember Jonestown. They see lives thrown away and they want something done about it. Those who hire "deprogrammers" face criminal penalties for kidnapping in some cases. Should that be so?

A once vast and stable consensus on the role of religion in America is slowly being chipped away. Everyone has a grievance: that he or she has been denied the right to worship, or that the government is stingy or stupid or petty, or that the government is favoring some religion other than his or her own, or that some religious leader has stuck his or her nose into business that is not his or hers.

How much religious liberty can we afford? And how much, if any, can we afford to throw away?

GOD

I do not want to wreck whatever good there is in this book by trying to write about God. Those who have done so best are either very wise and talented or very humble, or both. Spirituality that is full and overflowing cannot be counterfeited, and all the excellent writing about God that I have come across bursts with this spirituality. If you presume to write about God, you run the risk of exposing to the public and to yourself your own spiritual poverty. Try it and you will see what I mean.

But the question of God cannot be omitted from a book that aims to prep high school students for college. No matter where you go to school, no matter who your friends are or what you are studying or whether you believe in God or deny the existence of any and all gods, you will spend an enormous amount of time at school discussing the question of whether God exists. Of all the predictions I make in these chapters, I am most certain about this one. Whether your position in these discussions is "yes," "no," "the question cannot be answered," or "the answer makes no difference"—the entire universe of coherent answers—

you will be part of many discussions about it and you will answer the question.

Many of the conversations on this subject that I had as an undergraduate remain as vivid to me as though they had occurred this morning. I arrived at school to discover that I had been paired with a roommate who was a devoted Southern Baptist who read the Bible literally in all of its statements. When we wrapped up college four years later, he was a self-declared atheist. In between, there were quite a few relentless debates on the question. One occurred during exams when we were sophomores. A third student wandered into our room during this discussion, listened for a minute, and threw up his hands declaring, "Jesus Christ, it's finals and they're talking about cosmic bullshit!"

The combination of late hours and lots of free time lends itself to the capacity of students to debate the question of whether God exists for hours on end. Also, many personal convictions hinge on the individual's belief or non-belief in God. God comes in the back door of many discussions.

To present even a quick picture of the arguments you are likely to encounter would take me into the hazardous waters I discussed above, so I will not do that. I can, however, offer a few thoughts on the importance of arriving at an answer to the question, and of avoiding two of the four possible answers: "the question cannot be answered" and "the answer makes no difference."

Sooner or later you will hear of Pascal's "Wager." Pascal was a French doctor and philosopher whose book, *Pensees,* provides great reading in bits. Bits are all that I have ever read of it. The section on the "Wager" is somewhat long, complicated, and difficult to compress. Having alerted you to the fact that my recounting of the "Wager" may be flawed, let me tell you its basic logic as it has been explained to me:

If God exists and punishes unbelief, you will be punished in the hereafter if you do not believe.
If God does not exist, there is no reward for belief but neither is there a penalty.

The great gamble is persisting in unbelief when that unbelief may carry a great penalty while belief, even if wrong, carries no penalty and may in fact provide a great reward. Why not wager that God exists and simply believe?

Why not indeed? Many people would argue that belief is not cost free, that it requires modifications in personal behavior that rob the believer of a great many things that are to be valued, and may in fact twist the believer's personality and make him blind to truth. It is a gamble to believe, and it is a gamble not to believe, the rebuttal runs, so you must make a choice.

I have never been comfortable with the great gamble approach to God. God either does or does not exist, and whether He does or does not exist carries enormous consequences for individuals. It is a question that deserves a thoughtful answer, and not one calculated on risks and cost-benefit analysis.

Because the question can carry such enormous consequences, it not only deserves an answer, it deserves a serious inquiry.

Consider some other questions that deserve not only answers but also serious inquiry. Every individual approaching the decision "Should I marry so-and-so" confronts such a question. Not everyone gives the question a serious inquiry, but most people do and I believe all people ought to. Or take the question of accepting a job. Again, another critical decision, and depending upon how long one intends to stay employed in that particular job, a very serious review of the circumstances appears to be in order before saying yes or no.

Most serious questions do get the lengthy and thorough review they deserve. But the question of whether God exists is often excluded from this category, at least it was among many undergraduates of my day.

When faced with this question many folks are content to shoot from the hip, to improvise, or to rely on the answers of others or on habit. The arguments about God that take place in dorm rooms are among the most lazy, silly, and anti-intellectual I have ever heard. People who are in the "yes" camp often passionately repeat and repeat and repeat their own conviction, as if on this particular question their own authority should suffice. Some of these folks would never presume to answer other serious questions with a claim to authority, but on this topic they are bold.

The nay-sayers are equally astonishing in their rebuttals. With the same degree of presumption, often without any significant experience and without the benefit of crucial reading or exposure to learned teachers of the opposite view, they will still hold forth, quietly or loudly asserting that there is no proof, can be no proof, and that absent the possibility of proof, there is no rational way to answer yes.

The arrogance that surfaces on all sides during these discussions startles me in retrospect, even though I was as guilty as the rest. Also astonishing was the fact that on a question that everyone agreed was of great consequence, no one ever took the initiative to organize a coherent inquiry. It is doubtful that even one college offers a course "Does God Exist 101," but there is no reason that interested individuals cannot put one together. It is no small irony that students who spend hours and days plotting and planning a semester's course schedule never move beyond the status of couch potato when it comes to the question of whether God exists.

What would a serious inquiry involve? There are, no

doubt, a thousand variations, but some approaches are obvious.

Basic texts could be examined. Many believers and non-believers, for example, know very little about Scripture. The Gospels and the first five books of the Old Testament take only four hours to polish off. That is a good place to start.

Some will reply that this is dull reading, or too time-consuming. Fair enough, there are alternative approaches. Every paper in America has in its Saturday edition a listing of services at local churches. Nothing is more common to campuses on a Sunday morning—other than hangovers—than nothing to do. So make the rounds. Sermon-shop. Do not limit yourself to mainstream Christian denominations, though I do advise an acquaintance with the mainstream before you venture from the center. And, perhaps most important, engage some authorities—ministers, theologians, professors of philosophy who reject the idea of God—in conversation. I have never known one of these authorities to reject an invitation to dine with students.

Any one of these steps will have the effect of moving your dorm-room discussions away from windy generalities and into the realm of specifics. I do not suggest that the arguments end. Quite the contrary. As a Christian, I am convinced of the truth of the insight revealed in the first chapter of C.S. Lewis' *The Screwtape Letters* concerning argument. If you are not acquainted with this little book, you will want to read it if only for the fine writing and humor found therein. It is in the form of a series of letters written from an older, experienced devil to a younger tempter operating in wartime England. The older spirit writes the younger on the subject of arguments such as those I have been describing, (refering to God as "the Enemy" and an ordinary man as "the patient"):

> The trouble about argument is that it moves the whole struggle onto the Enemy's ground. He can argue too; whereas in really practical propaganda of the kind I am suggesting He has been shown for centuries to be greatly the inferior of Our Father Below. By the very act of arguing, you awake the patient's reason: and once it is awake, who can foresee the result? Even if a particular train of thought can be twisted so as to end in our favor, you will find that you have been strengthening in your patient the fatal habit of attending to universal issues and withdrawing his attention from the stream of immediate sense experiences. Your business is to fix his attention on the stream. Teach him to call it "real life" and don't let him ask what he means by real.

Anyone who involves himself in an earnest argument quickly comes to recognize the weakest part of his logic. As a lawyer, I have written hundreds of memoranda and briefs on legal disputes turning on questions of interpretation, and in every one of them there are flaws or gaps in the logic. If one side of an argument was perfectly airtight, there would be no dispute to begin with. The trick to prevailing in any argument is to force the opponent onto your ground, to position him so that he is attacking your logic where it is most sound while you detect and begin to pound away on the weakest link in his or her reasoning.

Lewis noted that arguments of any kind about the existence of God imperil the non-belief of non-believers. This happens, I think, because the non-believers quickly sense the holes in their own logic. The atheist who prefers to stay that way is best advised to shun the whole business of whether God exists. Paradoxically, these arguments only infrequently imperil the belief of believers, as logic is never for them the complete approach to the question; where reason stumbles, faith provides a footing.

The more learned the argument—the more steeped in

the basic texts, theology, and authorities—the more precarious the position of unbelievers. As a believer who would like to see more people recognize God, I advocate the serious study of the question and frequent argument on its merits. No hidden agenda here.

No one loses anything from the serious study of the question of whether or not God exists. Even the atheist who remains resolute in non-belief would carry away from it a literary and cultural education that will profit him immensely. Read St. Augustine's *Confessions,* or absorb the preaching techniques of an effective minister, or experience the Jewish Seder, and you will carry away, at the very least, knowledge of a great work of literature, pointers on effective rhetoric, or a lesson in the history that binds a unique people and which makes more accessible their collective experience.

All that is required is a little directed curiosity and a small amount of initiative. There is no more important question than whether God exists, and no question that more easily lends itself to painless but effective research. The four years ahead offer a unique opportunity to take up the study of the question and to use the abundant resources at your fingertips.

CONCLUSION

"I suggest," wrote E. M. Forster, "that the only books that influence us are those for which we are ready, and which have gone a little farther down our own particular path than we have yet got ourselves." He added: "You are being influenced when you say, 'I might have written that myself if I hadn't been so busy.'"

I have never read any of E. M. Forster's novels. I came across that quotation in a piece of literary criticism written by Joseph Epstein, a writer who always entertains and enlightens me. I read everything by Epstein that comes my way, even if his subject is far removed from my normal interests. Epstein is, for me, a habit.

Which brings us to the point of this conclusion: What is the next step? How do you inform yourself further about the many crucial issues that merit your attention, and how do you make that process not only balanced, but fun?

College will not help you much. Professors tend to omit the "balance" part frequently and the "fun" part almost always. At bottom, most of them want you to believe what they believe. But it can be done, and at a cost of under $150.00 a year.

One hundred and fifty dollars is not the cost of my subscription newsletter. It is rather the cost of a year's subscription to four magazines: *The American Spectator, Commentary, The New Republic,* and *The Washington Monthly.* These four journals cover the entire relevant agenda of American debate on most crucial issues. They present great writing. Subscribe. Read. Develop a habit.

I am proposing that you go in harm's way. Ideas and debate can be harmful. You could end up in the Jonestown of the 1990's if some screwball notion takes up residence in your head and you become obsessed with it. But if you let the full river of American opinion-writing flow through your mind, this danger is minimized and you greatly improve the chance that you will develop into a competent and critical thinker.

Magazines can be the most seductive of things. I binged on them for a long time, especially during law school when the course work was almost invariably dull. My pile of journals always beckoned me, and a good deal of enlightenment and entertainment could be had in less than 20 minutes. Articles in magazines rarely demand an extended attention span. Nicely segmented and usually full of variety, they present you with a fresh menu of intellectual fare at regular intervals.

They do, however, pile up and you may become attached to them. If so, you will soon be lugging boxes of back numbers from dorm to apartment.

But they will inform you and make you more learned than you had imagined possible. And possibly witty. And possibly a competent writer.

There is my radical approach. Four magazine subscriptions and the willingness to dip into them when they arrive. But this approach may not have occurred to you before, except in the context of *Sports Illustrated, Time,* and *Cosmo.*

After a year you should be able to whittle the four down

to three. Either *The American Spectator* or *The Washington Monthly* will present an editorial view you find quite incomprehensible. An additional cut will follow your decision to stand resolutely in the middle or to shade your information flow to one side, Left or Right. You may limit yourself to *Commentary* and *The New Republic,* organs of the neo-conservatives and the moderate Democrats respectively. But by then, if the taste for journals of opinion has been acquired, new titles will spring up on your shelves. The need for reasoned analysis is unrelenting once it is established. The habit becomes an addiction.

Reading along these lines will not turn you into a social dwarf. Confronted with an *SI* with Bernie Kosar on the cover and a *Commentary* that leads with British historian Paul Johnson on South Africa, you will still reach for the former, especially if you hail from somewhere near Cleveland. But later you will invest some time, about 30 minutes, in Johnson's piece and you will, as a result, be by far the most enlightened student on your campus concerning the complexity of that nation's agony.

As for books, I make no recommendations beyond the titles already put forward in the chapters, though it merits repeating that Solzhenitsyn's and Bukovsky's efforts possess the potential to change you completely and for the better if you take their message to heart. But generally, as to books, you are on your own. Course work in college will absorb too much time with reading lists to make additional suggestions. Make your own draft choices and do your own cuts.

This book is quite short. My publisher and I decided that length might scare off too many among the curious. Many, many things have been left out. But if you have made it this far, notice has been served. The intellectual atmosphere of college can be poisonous because so many different people want your adherence and your devotion and not merely your attention for a couple of hours a week. The antidote is freely available, and it is independent thinking and study. Even at the age of 18 you can duel with any professor; and if you walk away feeling foolish, it is not because you are stupid or limited. You just have to assemble the crucial facts and craft those facts into convincing arguments and counter-arguments. But you do have plenty of time ahead. Set some aside for the big questions.